m&m's

Around the World

An Unauthorized Collector's Guide

Ken Clee & Joyce Losonsky

Schiffer Publishing Ltd

4880 Lower Valley Road, Atglen, PA 19310 USA

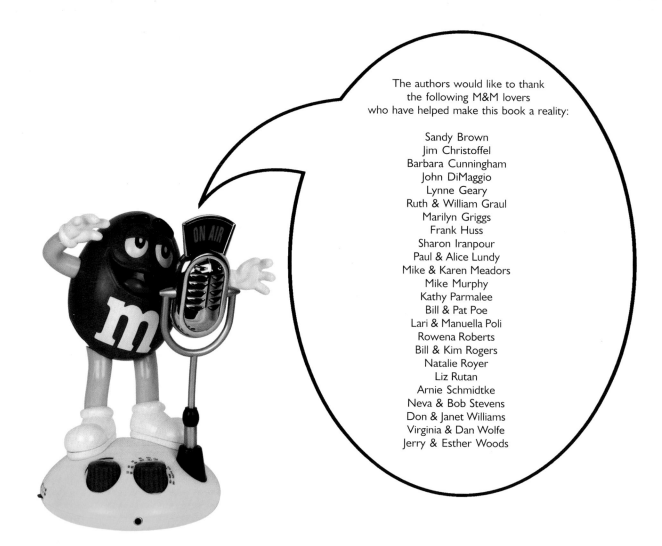

The authors would like to thank
the following M&M lovers
who have helped make this book a reality:

Sandy Brown
Jim Christoffel
Barbara Cunningham
John DiMaggio
Lynne Geary
Ruth & William Graul
Marilyn Griggs
Frank Huss
Sharon Iranpour
Paul & Alice Lundy
Mike & Karen Meadors
Mike Murphy
Kathy Parmalee
Bill & Pat Poe
Lari & Manuella Poli
Rowena Roberts
Bill & Kim Rogers
Natalie Royer
Liz Rutan
Arnie Schmidtke
Neva & Bob Stevens
Don & Janet Williams
Virginia & Dan Wolfe
Jerry & Esther Woods

Copyright © 2000 by Ken Clee and Joyce Losonsky
Library of Congress Catalog Card Number: 99-68560

Book Design by Anne Davidsen
Type set in Geometric 231 Heavy /Humanist 521

ISBN: 0-7643-1078-X

Printed in China
1 2 3 4

Published by Schiffer Publishing Ltd.
4880 Lower Valley Road
Atglen, PA 19310
Phone: (610) 593-1777; Fax: (610) 593-2002
E-mail: Schifferbk@aol.com
Please visit our web site catalog at
www.schifferbooks.com

This book may be purchased from the publisher.
Include $3.95 for shipping.
Please try your bookstore first.
We are interested in hearing from authors
with book ideas on related subjects.
You may write for a free catalog.

In Europe, Schiffer books are distributed by:
Bushwood Books
6 Marksbury Ave.
Kew Gardens
Surrey TW9 4JF
England
Phone: 44 (0)208 392-8585
Fax: 44 (0)208 392-9876
E-mail: Bushwd@aol.com
Free postage in the UK. Europe: air mail at cost.
Try your bookstore first.

Contents

Mars, Inc. Chronology:
A Kettle, a Bag of Sugar...and a Dream!

The history of the Mars Company is a rich blend of many compelling ingredients: people, places, and things. People such as Forrest Mars Sr. and Forrest Mars Jr.; places like Tacoma, Washington, Newark and Hackettstown, New Jersey, and Las Vegas, Nevada; and things like secret ingredients, mysterious operations involving the privately owned company, and marvelous marketing genius in designing great collectible characters and collectible items to promote the latest products. It is the following rich blend of ingredients which has developed the history of the Mars family into a candy dynasty.

1911 - Selling snack food was the primary business venture for the Mars family. Forrest Mars Sr., the candy salesman, and his wife Ethel, started their candy making operation on the West Coast in Tacoma, Washington. They had two sons, Forrest Jr. and John, plus a daughter, Jacqueline, all of whom provided leadership to the candy conglomerate through the years.

1932 - Forrest Mars, Jr. set up European operations in Slough, United Kingdom.

1930s - M&M's candies owe their origin to the Spanish Civil War when Forrest Mars, Sr. is said to have seen some Spanish soldiers eating pellets of chocolate that were encased in a hard sugary coating to prevent them from melting.

1940 - M&M Limited was founded in Newark, New Jersey by Forrest Mars, Jr. upon his return to the USA. He began making M&M's Plain Chocolate Candies and the history of the Mars dynasty was in the making!

1941 - M&M's Plain Chocolate Candies were introduced and became a favorite of American G.I.s serving in World War II. The M&M's candies were packaged in cardboard tubes and sold to the military as a convenient snack that traveled well. They were available in brown, green, red, and yellow.

1943 - Violet and orange were introduced into the M&M's color mix.

1948 - Packaging changed from tube form to the characteristic brown packet of M&M's Candies.

1949 - Violet color M&M's were replaced by the color tan.

1950 - M&M's Candies became a household word with the advent of TV in the 1950s. The little "m" imprint also began to appear at this time. Originally the little "m" was printed in a black color; this was changed in 1954 to a white "m" imprint.

1954 - Peanut M&M's were introduced as "the candy that melts in your mouth, not in your hand." They were designed to survive through the heat and humidity of an American summer—one without air conditioning. In the early '50s the advertising characters basically looked like stick people with clothing and an M&M shaped head. The letter "m" was very thin and the eyes of the characters appeared between the arches of the "m" imprint on the chocolate covered candies.

1957 - The advertising characters' bodies were dropped in favor of an M&M with white, pencil thick arms and legs. The letter "m" imprint on the candies was made bolder and the eyes stayed between the "m" arches.

1960s - M&M's Peanut Chocolate Candies added three new colors to the brown mixture: red, green, and yellow.

The famous 1954 M&M's slogan: "The milk chocolate melts in your mouth, not in your hands."

1955 M&M's chocolate wafer bar.

1962 - The "m" on the advertising characters became smaller and the faces appeared above the "m" for the first time while the feet of the characters were fattened.

1964 - Forrest Mars, Jr. merged his operations, which included the M&M brand candy operations, with his father's operations, which included the Milky Way Bar.

1970s - Father and son joint operations promoted the Snickers Bar in addition to the Milky Way Bar and M&M's Candies. This was a dynamic trio of candies.

1971 - The M&M's advertising characters became colorful and animated in advertising. The body of the peanut M&M was larger on the top than n the bottom (the reverse of the 1990s Peanut) and they had white arms and legs for animation ads.

1972 - M&M's colorful characters appeared on all packaging—reinforcing brand awareness.

1976 - M&M's Peanut Chocolate Candies added the color orange to the mixture. The red M&M's Chocolate Candies were removed from the traditional color blend due to controversy over a certain red food coloring (which was never used in the manufacture of M&M's candy).

1981 - M&M's Brand Candies expanded into the European market.

1983 - Australia and The Netherlands introduced Mars Candies.

1985 - M&M's Chocolate Candies introduced a holiday line of M&M products, using red and green for the Christmas season.

1986 - Pastel color M&M's in pink, blue, yellow, and green were introduced for the Easter season.

1986 -The advertising characters appeared with sneakers and gloves and fresh colored arms and legs. Austria, Belgium, Canada, Denmark, Ireland, Italy, Spain, and Switzerland introduced Mars Candies.

1987 - Red M&M's were reintroduced to the standard mix of candies after an eleven-year absence. France, Germany, Hong Kong, Japan, and the United Kingdom introduced Mars Candies.

1988 - M&M's Candy toppers appeared on the market, along with the introduction of the almond variety of M&M's for the holidays.

1989 - M&M's plain and peanut mint variety became available for the holidays. Malaysia introduced Mars candies.

1990 - The advertising characters took on a "modern look," with white nondescript sneakers and animated facial expressions. The peanut butter variety of M&M's plain and peanut was introduced, as were the Valentine colors of white, pink, and red. Mars continued to introduce their products throughout South America, Europe, and Asia.

1992 - Seasonal candy toppers appeared with the Easter season and almond variety M&M's plain and peanut became available year round.

1993 - Summer and Winter Olympics events were portrayed on M&M's candy toppers.

1995 - Mars asked consumers to vote for a new color. The choices included blue, pink, purple, or no change. BLUE won by a landslide! The color tan was retired and Mars Chocolate Baking Bits were added to the product line.

1996 - M&M's Minis Milk Chocolate Candies were introduced in six different colors: red, blue, green, yellow, orange, and brown. They were sold in recloseable plastic tubes. M&M's Studio Brand Characters appeared on the Internet with the opening of the "M&M's" Studios website. The Halloween colors of dark brown and orange were introduced.

1997 - M&M introduces the newest character—GREEN. The first female M&M character, Green joined her male colleagues—Red, Yellow, and Blue. She is a '90s woman, and toured the USA proclaiming, "I melt For No One!" The M&M characters were redesigned with more expressive faces as well as eyes that were not flat on the character's face. Each character began to take on an individual personality in the advertising. Chocolate World opened in Las Vegas, and includes the M&M's World retail store above Ethel's Chocolates (an upscale Mars chocolate retail store).

The color orange was added to M&M's Peanut Chocolate Candies in 1976.

M&M's are an international favorite!

Red was discontinued from the traditional color blend in 1976 but brought back in 1987.

M&M toppers, now one of the most popular collectibles, first appeared atop tubes of candy in 1988.

1998 - M&M advertised itself as "the candy of the NEW MILLENNIUM and beyond," based on "MM meaning Year 2000." The name was a perfect fit! Mars painted their NASCAR as an M&M's car, and it was an immediate winner!

1999 - A confetti-colored mix for the New Millennium was introduced consisting of white, red, yellow, purple, green, and light blue. M&M' Crispy Chocolate Candies were introduced with the Orange Crispy character leading the advertising appeal. Forrest Mars Sr., architect and patriarch of the international candy and food conglomerate, died at the age of 95. Ernie Irvan, driver of the successful M&M's car, retired.

2000 - Mars products are enjoyed in over one hundred countries throughout five continents!

Blue—introduced in 1995 as M&M's newest color!

M&M's Minis Milk Chocolate Candies hit the market in 1996.

Pricing and Values

Anyone who has ever written a price guide has faced the challenging task of placing values on items included in the book, and later heard from dealers that the prices shown were too low and from collectors that the prices shown were too high.

Part of the problem can be that items available only in certain parts of the world vary in price depending on how far away one is from where the item was originally distributed. In young hobbies, there often have not been enough transactions on some items to determine their true value.

Prices are sometimes inflated when a previously unknown item appears for the first time. Most hobbies have a few collectors who must be among the first on the block to get any newly surfaced item, while the vast majority of collectors are willing to wait until those few pay a premium to get it first. Usually, the price then comes down a bit, especially once the item becomes readily available. (There are always exceptions, of course, where only a few of any one item ever surface and the price stays high or even rises.) As an example, in the spring of 1998, European 7" dispensers were selling regularly on Internet auctions at well above $100 each. As more people made contact with their friends in Europe and the dispensers became easier to obtain, the price slowly dropped to its current value, easily 75% lower than the early price. In placing values on such items, therefore, both extremes must be included to arrive at an appropriate range of value.

Generally speaking, most collectibles are more valuable if they are still mint in their original packages. The better the condition, the higher the value.

Additionally, you should expect to pay a higher price if you buy an item from a dealer. The dealer has either done your hunting for you, or has paid a collector or another dealer to obtain the item and expects to make a profit for his time and/or trouble. If you are willing to do the hunting yourself, you will usually find treasures at a much better price than if you pay a dealer. However, you must have the time, enjoy the hunt, and understand that you may never find the item you seek yourself. If you really want the item, or if you don't have the time or desire to search for it yourself, then paying a premium price to a dealer is certainly a viable option. Besides, you may get an excellent deal from the dealer,

M&M Advertising for "The Official Candy of the New Millennium."

depending on how much it cost the dealer to obtain the item, and how anxious he or she is to complete a deal on that particular day.

Prices vary for so many reasons that the prices in a price guide are only reliable for helping you judge the actual price you may see for an item. Perhaps that is why they are called price "guides" and not price lists. The actual value of any item on any day is purely a result of how much a dealer is willing to accept and how much a buyer is willing to pay. If those two factors meet or overlap, a price is established and a sale takes place. However, that price represents the value of the item for those two people at that particular point in time. Two other people could exchange the very same item on the very same day at a price similar or vastly different from the first transaction.

Collectors vary in their attitude towards prices as well. Some collectors have a very definite limit on what they will pay for an item at a particular time. They frequently pass on an item because it is a few dollars more than the figure they have in mind, or because the dealer would not offer any discount, etc. However, others believe that when they factor in the time it took to get to a show that day, the cost of gas and depreciation on their car to drive there, the time spent looking for the item, and perhaps the cost of getting into the show, turning down any item for a dollar or two is equivalent to straining gnats and swallowing elephants. If it has already cost six hours of your time and $50 in gas,

7

tolls, depreciation, entrance fees, and parking, why go home with nothing? Still other collectors are interested in an item only if they believe they are paying less for it than everyone else. This is surely what makes the world go 'round! It's also what makes some people see prices in a price guide as too high while others see them as too low.

There is also a difference in the expectations that people have for price guides. Some think the prices in a guide should reflect what they could expect to receive if they were to sell their collection—or parts of it—to someone else. Others want a guide that tells them what they can expect to pay for new items to add to their collection. Obviously, these two numbers cannot be the same, or there would be no profit for a dealer, toy store, antique mall, or occasional flea market seller, and all would quickly disappear. Price guides sometimes reflect seller numbers and sometime reflect buyer numbers, while many try to straddle both figures with ranges that represent both numbers.

In actuality, most dealers are not interested in buying things unless they feel they can sell them at 50%-200% more than they paid for the item or collection. It may not sound fair to the collector, but it is reality. The dealer who is selling an item for $5 probably would not pay more than $1-3 for the same item.

Our approach to pricing is to publish a guide that reflects the range of prices collectors should expect to pay for items in very good to new condition at a show or from a dealer, shop, or antique mall. You may find some items at prices below the range if you are willing to hunt, willing to accept the item in lesser condition, or just get lucky. On the other hand, you may find an item at a higher price because the only person who has the item of your dreams simply won't sell it at a lower price. Or you may never find it at any price.

Generally speaking, if you are looking to sell your items to a dealer or shop as a collection or to sell items quickly to raise cash, you should expect to receive only a third to a half of the values shown here for the items. If you sell items individually, or over time to other collectors, then you may certainly get the values shown in this guide.

Like many other guides, this book offers listings and pictures of extremely rare, unusual, and seldom seen items. Obviously, there are no established values for these items. Rather than just slap numbers on them, we have chosen to show either no value or a range of values you might expect to pay based on the previous sale of items similar in appearance, vintage, scarcity, or collectibility.

These prices also reflect what we believe to be current values. As time passes, these prices will change slightly or greatly, based on the items' availability, desirability, and the rate of growth or decline in the number of people in the hobby. Prices or values for categories like toppers, dispensers, tins, Pochette Surprise toys, and the like are pretty well established due to their popularity, availability, and history of prices realized in a great number of previous sales.

We hope the above explanation provides some insight into why some people see price guide prices as too high while others view those same prices as too low, and hope that our efforts are truly helpful to the vast majority of hobbyists who have purchased this book. Just realize, please, that they are guidelines only, and that the value of any item is truly established between two individuals at a time and place when the transaction actually is completed.

Where to Find M&M Items

One factor affecting the value of M&M items is their availability. With a little imagination and some leg work, you can add unique items to your collection. Here are some of the places you can find M&M items.

The Basics

• Retail Outlets - Items can be found in major toy store chains. Stores such as Nieman-Marcus, Wal-Mart, and Target have had promotions for items not available anywhere else. Party supply stores can be another good source.

• Mail-In Offers - Always check M&M's packages in your grocery store for special items available directly from MARS. Normally, these offers require a payment and one or more proofs of purchase.

• Dealers, Flea Markets, and Antique Shows and Shops - Don't be afraid to dig deep in boxes under tables and on the back of a truck at a flea market. Ask dealers at shows. They may have just what you are looking for and be more than happy to bring it with them next time. Find mail order dealers and get on their mailing list.

• NASCAR Stores, Races, or Shows - M&M's racing team memorabilia, especially shirts, hats, jackets and die-cast cars, is available for sale.

• The M&M Collector's Club - The club offers a web page with all kinds of great information on where to find old and new items, upcoming major shows, and classified ads from members. They also hold an annual convention for members and publish quarterly newsletters. Their website address is http://www.mnmclub.com or contact the authors for further information. This is a collector's club and is not affiliated with M&M/Mars.

The M&M/Mars Home Page - The MARS Corporation maintains a website at www.m-ms.com that offers information, fun and games, and the opportunity to purchase a few new items. You can join their "Inner Circle"™ and purchase items available only to members. You can also join the NASCAR/M&M's Car fan club from the same website.

Advanced Methods

M&M's World™ Store in Las Vegas - You will never see more M&M items in one place at one time. Call them to get a copy of their mail order catalog.

Store Displays - Sometimes a friendly store manager will agree to let you have certain display items at the end of a promotion. This often requires several tries before success is achieved.

Airport International Departure Lounges - Several very desirable M&M collectibles have been available only at the duty-free counters found in the international departure lounges of airports. You have to be leaving the country to buy them. If you know of someone coming for a visit from another country, you might ask them to pick up something for you at the airport.

Trading Partners in Other Countries - Locate collectors in other countries and trade items with them, or find someone willing to shop for you at flea markets and stores in other countries. Friends vacationing outside your country can sometimes be convinced to pick up things for you as well.

Employee Only Items - Employees/retirees of MARS can order from their very own catalog of items not available to the rest of us. They are discouraged from buying items for any use other than their own, so please don't ask them to get things for you. However, you can certainly apply for a position with MARS, or marry a current employee!

Additions and Corrections

To the best of our knowledge, the information in this book is accurate and timely as of this writing. It is our first edition, and there may be small (we hope) errors and certainly some omissions that we can include in the next edition. We invite your ideas, additions, corrections, and encouragement for future editions. No book on this subject can ever be complete, as both old and new items surface every day. We believe our listings of toppers, dispensers, and pocket surprises are up-to-date and complete, but have not attempted to show every item in the M&M universe. Although we show a sprinkling of items from the M&M's World™ and Inner Circle™ groups that we think are exceptionally nice or potentially collectible in the future, we feel that most readers can order their catalogs and see all the items they have to offer. Our goal is to show the more collectible, potentially collectible, and obscure items that are out there. With the universe shrinking every day and items from around the world becoming easier to find and acquire, we believe the collecting of M&M items from one country only is robbing a collector of terrific items that are easily obtainable.

The authors are collectors as well as writers, and invite your inquiries if you have items for sale or are looking for items to buy. You can contact them by mail, e-mail, or phone as follows:

Ken Clee
P.O. Box 11412
Philadelphia, PA 19111-0412
(215) 722-1979
E-mail: waxntoys@aol.com

Joyce Losonsky
7506 Summer Leave Lane
Columbia, MD 21046-2455
(410) 381-3358
E-mail: JoyceUSA@aol.com

Alphabetical Photo Gallery

Advertising

1998 ad for "Free M&M's with purchase of a collectible 32 oz. Fountain Drink." Three different cups were offered by Ultramar's Diamond Shamrock Gas Stations in the Southwest USA.

Above: Advertising items from Mr. Bean's European promotional tie-in included this 2-sided postcard-sized ad from Germany and candy package from Europe. Postcard $10-20

Left: Halloween advertising balloon. $5-10

Miscellaneous promotional advertising includes posters of all sizes.

This 6" by 7 1/2" pennant is from a string of eight identical European pennants displaying promotional advertising for the introduction of the blue M&M. $15-20

998 Inflatable vinyl advertising blimp, measures over two feet long. his photo shows a mint-in-package blimp. $25-50

1999 Inflatable vinyl bag of Crispy candy with the orange character hanging below it, mint in original package. Loose $15-25; MIP $25-50

This two foot tall, inflatable vinyl blow-up from Germany certainly caught the eye of shoppers. It is referred to as the Janet Jackson balloon. $40-75

Awards

From left: Employee Award brass medal given for excellent performance at different jobs within the factory; Employee Recognition brass medallion on wooden stand for responsibility, mutuality, freedom, efficiency, and quality on the job; brass medallion for recognition of employee's contribution to Mars' Global Campaigns. Each $25-50

Salesman's promotional items: "Ten Years in the Heart of Texas (Waco)" brass medallion, leather wallet, 1987 date book, and M&M/Mars etched glass.

M&M/Mars Silver Anniversary Coin (1958-1983), commemorating twenty-five years in business. $20-40

Above: 50th Birthday of M&M's Chocolate Candies brass coin (1946-1996). $20-40

Center Right: 1975 pewter plate presented to salesmen for attainment of sales objectives. 9 1/2" dia. $20-40

Right: 1981 pewter plate presented to salesmen for "One Billion in Sales," 9 1/2" dia.; desk set with medallion. $20-40

Bags

1990 Tote bag.

Miscellaneous vinyl and cloth tote bags and school bags
in varying sizes, colors, and shapes.

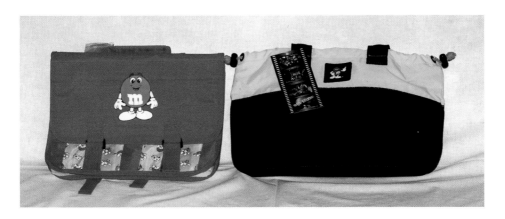

15

Assorted canvas bags and totes.

Braves backpack and Kroger lunch bag. Each $30-60

These one foot tall plush backpacks were available in four colors.
The tags stated, "Welcome to the World of M&M's!" Each $20-30

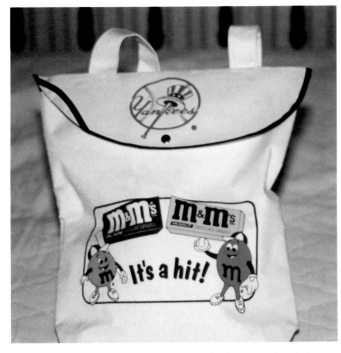

Yankees backpack. Notice the boxes of M&M's pictured and the character's skinny legs. $30-60

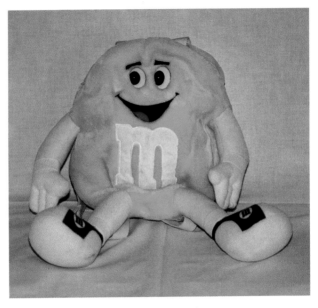

1990s Asian plush character backpack. $20-40

Banks

Beach Items

Plastic M&M's bank. Rare

Beach towel. Opposite side has the brown M&M's color. $20-40

1998 Small golf towel, available from Employee's Catalogue. $20-25

Large round terrycloth beach towel. $20-40

Bedding

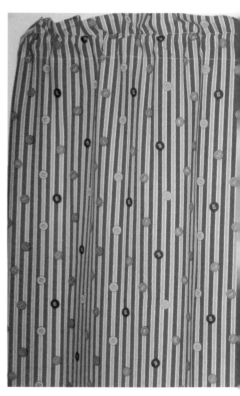

Red and white striped twin sheets, pillow, drapes, and comforter set.

Cannon Cowboy pattern comforter. MIP $50-100

Cannon Cowboy pattern twin sheets and pillowcase set.
MIP $30-50

Bendies

M&M Bendy figures can be found in all four colors (green, blue, red, and yellow) and in various sizes by themselves and/or within promotional packages.

Assorted small size (2" - 2 1/2")
bendies. Each $4-5

Assorted medium size
(3 1/2" - 4") bendies. Each $6-12

Above and top left: Assorted large size (6"-7") bendies, shown out of the package and mint in the package. Each MIP $15-25

6" Plastic sand pail with large bendy and candy packages sold in retail stores during summer 1999. Pails came in four colors: green, blue, red, and yellow. Each $10-20

Books

Counting Books: 1994 *Counting Book* (8 1/2" x 11" with orange M&M for the "o" in "Counting." A later version of the *Counting Book* (also dated 1994) has blue M&M for the "o" in "Counting." Also pictured is the *Counting Board Book* version (6" x 6", dated 1997) and the 1998 *Math* book. Prices range from $15 for more recent editions to $30 for older editions.

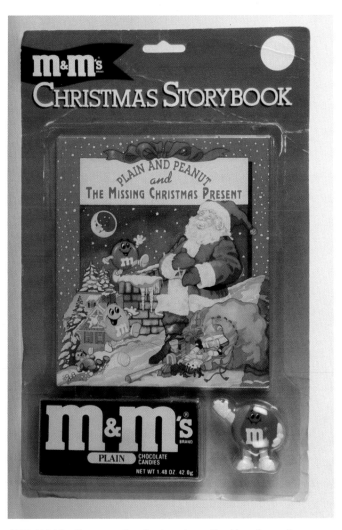

1993 *M&M's Christmas Storybook* with small rubber figure (bottom right) and unusual box of M&M's (bottom left). Most M&M's have been packaged in bags throughout the years. Book alone $5-10; Book on card $15-25

Sportlight Coloring & Activity Book (8" x 10 1/2"), offered in conjunction with M&M's sponsorship of the 1984 Olympic Games. Rare.

The M&M's Characters Fun Coloring Book. $20-30

Published in 1997, *A Little Encyclopedia of M&M/MARS* (6 1/4" x 9") contains twenty pages of information on the company and its products. $15-20

Booklets - Recipes

Red's Favorite Recipes
FEATURING
CHOCOLATE Mini Baking Bits

Brighter Baking With M&M's Chocolate Mini Baking Bits from 1996 was available in either paperback (96 pages, 5" x 8") or hardback (8 1/2" x 11") versions. Hardback version $20-25; Paperback version $10-15

Red's Favorite Recipes booklet from 1997 measures 5 1/2" x 8 1/2" and contains sixteen pages of mouth watering snacks and desserts. $5-10

Boxes

Old 5¢ M&M's boxes! Note the thin legs on the old, original characters. $20-30

Buttons

This Special Olympics button is 2 1/2" in diameter. $5-10

M.A.D. button and 1998 promotional button. Each $5-10

M&M's World of Las Vegas, Nevada, 3 1/2" diameter button worn by employees, 1999. $5-10

Calculators

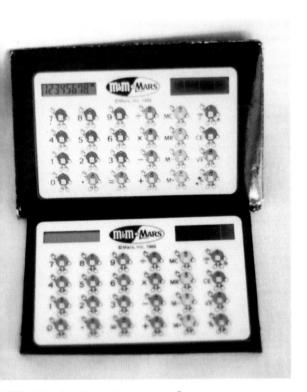

1989 White faced calculator in case. Rare

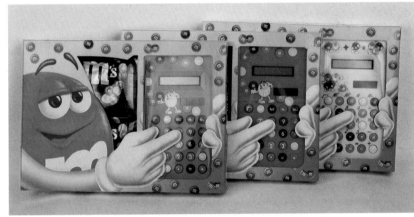

These calculators came in yellow, red, and blue. They were available with at least three different character designs on the calculators and at least three different packaging designs outside the US. They were also available with no packaging at the Las Vegas store. 3" x 5 1/4". Each loose $15-25; Each MIP $20-35

From "down under" comes this 1999 round plastic 4" diameter yellow solar calculator. $20-30

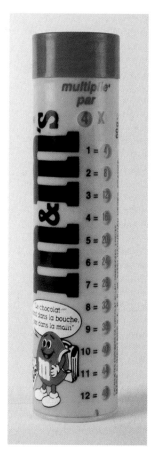

Not offered in the USA, these early 90's 7 1/2" hollow tube calculator/pencil cases came in at least two different colors (yellow and brown) and included multiplication and division tables. They also came in several designs and languages. Each $15-30

Calendars

M&M/Mars started 1988 off with this Fun Baking Calendar, which included recipes and coupons. $25-40

Cameras

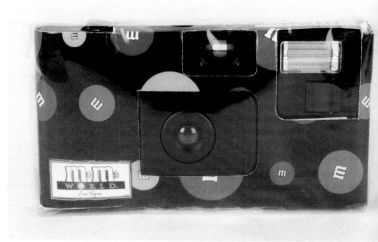

1999 M&M's World in Las Vegas sold this black cased camera with cardboard sleeve. $20-25

1990 "Smile" Camera was a mail-in premium and came with a case displaying the M&M's peanut bag on one side and the M&M's plain on the other. $25-40

Retail outlets sold these small, compact cameras in sets of three different colors, six colors in all. Set of 3: $30-50

Candy Jars and Dishes

M&M's candy dish from Holland.

Assorted crystal and glass candy dishes hold the chocolates pieces in style!

M&M's saluted The United States Air Force's 40th Anniversary with this candy jar in 1987. $40-60

1999 Plastic candy dish sold in retail outlets. $10-20

This tall plastic candy jar was sold on the QVC Shopping network in 1999. $40-60

CD's

1995 M&M's European "Musical Magic" CD comes pre-packaged with candy treats. $20-30

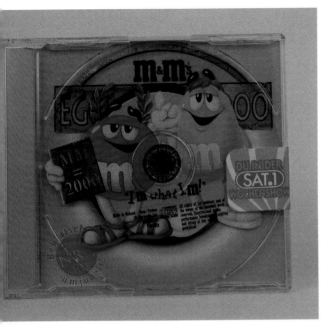

M&M's "Sings the Blues" on these Volume 1 and 2 CD's from Europe. Volume 2 is actually shaped like the character, making it a CD collector's favorite also. Volume 1 $20-30; Volume 2 $25-40

1999 "I M What I M"CD from Holland features the title track and three other songs. $20-30

Chocolate Chums

1987 was the year for Chocolate Chums! Pictured is a set of Chocolate Chum plush 14" Bears from Heartline, including the Snickers bear, M&M's tan (peanut) bear, Milky Way bear, and M&M's chocolate bear! Each $20-40

Chocolate Chum chocolate bears came thirty to a box, designed to satisfy the heartiest of chocolate attacks! They were sold in candy retail outlets. Rare

These 1987 female Chocolate Chum bears came all decked out in ribbons and bows! Standing just under 3" tall, flocked and gorgeous, these rare figures include a chocolate M&M's (plain) girl with a white bow in her hair, tan Snicker's girl with a blue ribbon around her neck, gray Milky Way girl with a white bow in her hair, and brown (peanut) M&M's girl with a white ribbon around her neck. A very small quantity of the female versions was manufactured. Each $15-25

The 1987 male Chocolate Chum flocked bears are not nearly as difficult to find as their female counterparts. They stand just under 3" tall and include four standard bears and one variation: gray Milky Way bear, Snicker's bear with hands out or Snicker's bear with hands behind back, M&M's tan (peanut) bear, and M&M's plain bear. Each $3-6.

These plastic 1987 Chocolate Chum refrigerator magnets stand 2 1/2" tall. They came in all four figures. Each $3-6

1987 Chocolate Chum plastic book clips are 4" tall and were also available in all four characters. Each $4-6

Christmas Items

Carl's Jr., a West Coast fast food restaurant chain, gave out this set of five-page heavy cardboard die-cut books measuring 5" x 4" in 1989. M&M characters appeared on each storybook page. The complete set included: *Timmy's Tree Trimmers*, *The Perfect Snowman*, *The One That Almost Got Away* and *A Happy Note*. Each $15-25

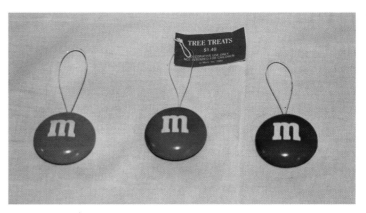

1988 Tree Treats came in three colors: orange, green, and red. Rare

Assorted M&M/Mars holiday ornaments add just the right touch to a holiday tree!

Some European holiday surprises in 1999 came in colorful M&M's decorated boxes!

Pictured is a 1997 set of six prototype, plastic, snap-out Christmas ornaments from Germany. These were never commercially produced and are considered extremely rare. They are marked "Not For Sale" on the outer packaging. Each loose $25-40; Each MIP $30-60

M&M's World in Las Vegas retailed this
6 1/4" tall Happy Holidays glass in 1998.
$10-15

These three 1998 Christmas tree ornaments from M&M's World in Las Vegas (the two figures in
the center are connected) are made of poly-resin and are approximately 3 1/2" tall. Each $10-15

M&M's World Las Vegas retailed this "Happy Holidays" ceramic plate
in 1998. $10-15

M&M's Secret Santa Snow Globe was available at USA retail
outlets during the 1998 Christmas holiday season. $8-12

Clocks

is 7" plastic Big Red alarm clock came from "down under" in 1999. 0-60

"The milk chocolate melts in your mouth - not in your hand" slogan appears on this 1979 plastic battery-operated School House style clock! Note the skinny legs on the M&M's characters. $50-100

low and Red in baking outfits are featured on this 1990s ker's clock! $50-100

"MM means 2000, The Official Candy of the New Millennium." This 5" plastic, battery-operated Millennium Countdown digital alarm clock counted down to zero at midnight on December 31st, 1999, then functioned as a real clock thereafter. $20-40

"New Blue" added just the right ring to this alarm clock, issued at Blue's introduction to M&M's parade of colors. $30-50

Clothing

This Crispy belt was available to employees only at the in-house gift shop or ordered through the Employees Only catalogue. $20-30

Above: Blue was the appropriate choice to decorate this Blue Jeans jacket, sold at retail stores, NASCAR stores, and at M&M's World, Las Vegas. $50-100

Right: This M&M's World scarf adds just the right sweet touch to a festive occasion. $15-25

M&M's Peanut Butter apron came in red for ready to bake! $20-40

1997 Simplicity Patterns for M&M's Halloween Costumes, in both adult and child sizes (Minis Bag). Each $5-15

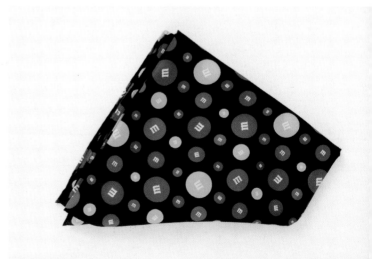

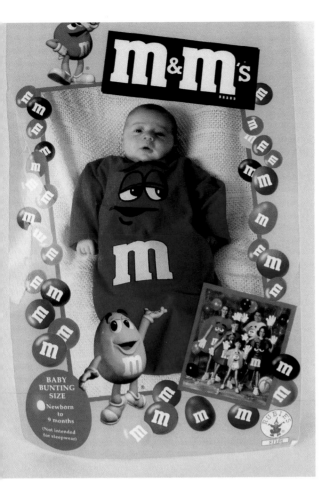

Baby Bunting and Halloween costumes (in Small Adults, Medium Adults and Large Size Adults) were sold in retail stores in the earlier '90s. Each came in different sizes and the four basic colors: red, yellow, green, or blue. Costumes featuring the new characters appeared in the late '90s. Each $10-30

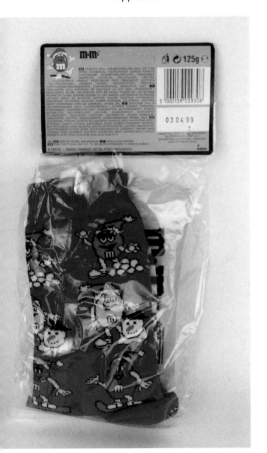

Sold in Germany, these M&M's socks were available in holiday and seasonal styles in a variety of sizes. $15-30

33

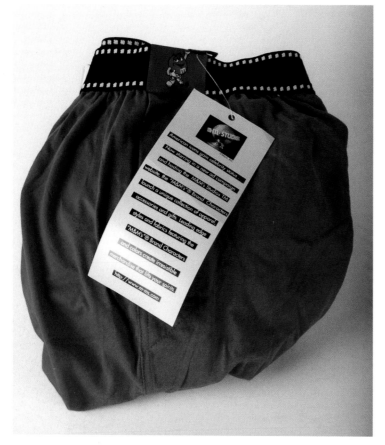

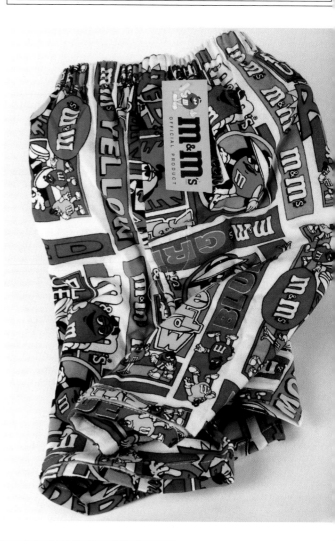

Shorts and boxers are available in many different patterns. Shown is an assortment of colors and designs. Each $15-25

M&M's Sunglasses were a mail-in offer in 1988. The carrying case was two tone: brown on one side and yellow on the other. A strap was included, covered with M&M characters. $15-25

34

Assorted T-Shirts abound! Even McDonald's selected
M&M's to advertise their latest ice cream venture—
McFlurry! Each $10-25

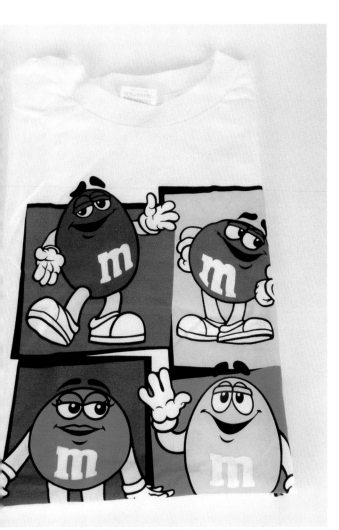

This M&M's Mars windbreaker jacket and design came in a selection of colors and character designs at one upscale retail store in the mid-1990s. $50-75

Cookie Jar

The slogans, "I love what I eat" and "MM = 2000" are included on these assorted ties. Many other styles also available. Each $20-25

1982 8" tall ceramic cookie jar is beige with a brown lid. $100-200

Cups

Assorted plastic cups come in various colors, designs, and styles throughout the world. Here is just a sampling.

Ultramar's Diamond Shamrock Gas Stations offered 32-ounce promotional plastic cups that came in three designs, and were numbered 1 of 3; 2 of 3; and 3 of 3. Set $15-25

M&M/Mars 1998 monstrous 9" tall Color Changing Cup was sold at movie theaters around the USA. When cold liquid was added, the design changed! $6-10

Atlanta Broadcasting Advertising Club, Inc. gave away this promotional can cooler in 1986. Rare

Dispensers

Dispensers are pure fun—they come in many different shapes and sizes. Confusion exists about the country of origin for some of the international dispensers.

3" DISPENSERS

All 3" dispensers have a slide window in the back for holding the candy. Open it to put the candy in, open it to take the candy out. The most common of the 3" dispensers are the ones that were made initially in both plain and peanut in brown, green, yellow, red, and orange. They feature the figure with one arm to his side and the other arm raised and waving. These seem to have been distributed all over the world. Recently, the same dispenser has surfaced in dark blue peanut from Australia and a powder blue peanut from Thailand.

There are some great variations to the 3" dispenser. In the mid 1990s, FTD had a promotion featuring a 3" yellow or red plain dispenser holding a bouquet of flowers in one hand. This dispenser came with matching color plastic pot with an "M" on it, as well as a gift card. This also seems to have been available all around the world. In addition, green plain figure (without matching pot) surfaced outside the US. Recently, a couple of orange variations and a couple of green ones with flower variations seem to have been offered in the Far East.

There were also peanut dispensers issued in Europe for the Easter holidays with the figure holding an Easter egg under his arm, and again at Christmas holding a sled under his arm or a Christmas stocking in one hand. Each was available in three or four color variations. These were not offered in the US.

" - USA

❏	Waving - Plain Orange	$4-8
❏	Waving - Plain Red	$4-8
❏	Waving - Plain Yellow	$4-8
❏	Waving - Plain Green	$4-8
❏	Waving - Plain Brown	$5-10
❏	Waving - Peanut Orange	$4-8
❏	Waving - Peanut Red	$4-8
❏	Waving - Peanut Yellow	$4-8
❏	Waving - Peanut Green	$4-8
❏	Waving - Peanut Brown	$5-10

3" - International

❏	Holding Easter egg - Peanut Yellow	$15-25
❏	Holding Easter egg - Peanut Red	$15-25
❏	Holding Easter egg - Peanut Green	$15-25

" - USA

❏	Holding flower bouquet (FTD) - Plain Red	$10-15
❏	Holding flower bouquet (FTD) - Plain Yellow	$10-15

3" - International

❏	Holding sled - Peanut Red	$25-40
❏	Holding sled - Peanut Yellow	$25-40
❏	Holding sled - Peanut Green	$25-40

" - International

❏	Waving - Peanut Blue (Light Blue) (Thailand)	$75-150
❏	Waving - Peanut Blue (Dark Blue) (Australia)	$30-45

3" - International

- ❏ Holding Christmas stocking - Plain Red $25-40
- ❏ Holding Christmas stocking - Plain Red $25-40
- ❏ Holding Christmas stocking - Plain Yellow $25-40
- ❏ Holding Christmas stocking - Plain Green $25-40

3" - International

- ❏ Holding green flower bouquet (FTD) - Plain Green $20-30
- ❏ Holding multi-color flower bouquet (FTD) - Plain Green $30-60
- ❏ Holding white flower bouquet (FTD) - Plain Green $30-60
- ❏ Holding multi-color flower bouquet (FTD) - Plain Orange $30-60
- ❏ Holding green flower bouquet (FTD) - Plain Orange $30-60

7" DISPENSERS

7" Dispensers were issued only outside the US and open just lik the 3" dispensers. Six poses were available; a collector would need t obtain twenty-two of the dispensers to cover all of the possible comb nations of figures and colors.

Note: The orange striped ski variation was found only in the re peanut skier with green skis. The skis on this dispenser are a single pla tic piece forming a right angle, as opposed to the more common variet with two separate skis joined together at a right angle.

7" - International (1991)

- ❏ Gift box (green) on foot - Plain Orange $30-50
- ❏ Gift box (green) on foot - Plain Red $15-30
- ❏ Gift box (green) on foot - Plain Yellow $15-30
- ❏ Gift box (red) on foot - Plain Yellow $15-30
- ❏ Gift box (red) on foot - Plain Green $15-30

7" - International (1991)

- ❏ Holding Easter egg - Peanut Red $15-30
- ❏ Holding Easter egg - Peanut Yellow $15-30
- ❏ Holding Easter egg - Peanut Green $15-30

7" - International (1991)
❏ Soccer player - Plain Red $15-30
❏ Soccer player - Plain Yellow $15-30
❏ Soccer player - Plain Green $15-30

7" - International (1991)
❏ Skier (yellow skis) - Peanut Red $15-30
❏ Skier (green skis) - Peanut Yellow $15-30
❏ Skier (red skis) - Peanut Green $15-30
❏ Skier (green skis) - Peanut Red (with white stripes on skis)
     $15-30

7" - International (1991)
❏ Wearing beret, holding Easter egg and paintbrush - Plain Red
     $15-30
❏ Wearing beret, holding Easter egg and paintbrush - Plain Yellow
     $15-30
❏ Wearing beret, holding Easter egg and paintbrush - Plain Green
     $15-30

❏ Skier (green skis) - Peanut Red (with orange stripes on skis)
     $35-60

7" - International (1991)
❏ Weightlifter - Peanut Red $15-30
❏ Weightlifter - Peanut Yellow $15-30
❏ Weightlifter - Peanut Green $15-30

9"-10" DISPENSERS

These terrific mechanical dispensers were offered first in the US at Christmas of 1992. A feature of recent holiday seasons has been the appearance of the new Holiday dispenser. After an initial offering of the plain figure in which the candy came out in one hand after pulling the other arm down, M&M offered a series of sports dispensers (football, baseball, basketball players), and most recently has offered the Sax player and the recliner dispenser.

Each of these figures was available in one color in the US, with a couple of exceptions, and in the same color and/or several other colors around the world. The dispensers issued inside and outside the US in the same color can only be differentiated by their packaging. Their value depends on how widely each was distributed.

9"-10" USA (1992)
- ❏ Dispenses into hand - Peanut Yellow $15-25
- ❏ Dispenses into hand - Plain Red $15-25

9"-10" USA (1993)
- ❏ Dispenses into hand - Peanut Red $15-25
- ❏ Dispenses into hand - Plain Green $15-25

9"-10" USA (1995)
- ❏ Football player - Peanut Red $15-25

9"-10" USA (1998)
- ❏ Football player - Peanut Green (M&M's World Store exclusive) $25-35

9"-10" USA (1998)
- [ ] Sax player - Peanut Blue $15-25

9"-10" USA (1996)
- [ ] Baseball player (blue hat) - Plain Orange $15-25

9"-10" USA (1999)
- [ ] Recliner - Peanut Yellow/red chair $15-25
- [ ] Recliner - Peanut Yellow/black chair (not pictured) $20-35

9"-10" USA (1997)
- [ ] Basketball player - Peanut Blue $15-25

9"-10" - International

- ❏ Dispenses into hand - Plain Blue $75-125
- ❏ Dispenses into hand - Plain Yellow (not pictured) $60-100

9"-10" - International

- ❏ Football player - Peanut Yellow $60-100

9"-10" - International

- ❏ Baseball player - (blue hat) - Plain Red $40-60
- ❏ Baseball player - (blue hat) - Plain Yellow $40-60
- ❏ Baseball player - (blue hat) - Plain Green $40-60
- ❏ Baseball player - (red hat) - Plain Yellow $40-60
- ❏ Baseball player - (yellow hat) - Plain Green $40-60

9"-10" - International

- ❏ Basketball player - Peanut Red $40-60
- ❏ Basketball player - Peanut Yellow $40-60
- ❏ Basketball Player - Peanut Green $40-60

Dispensers - Early, Mechanical and Miscellaneous

In the past few years, some very clever mechanical dispensers have been offered. Most notable are the Dilbert dispenser of 1998, and the orange Krispy "Hide'N'Hander" and "Jammin' Red" dispensers of 1999. There have also been a series of smaller dispensers issued, such as the Minis dispensers (in which the tubes were loaded in the back of a character or in a truck) and the Ernie Irvan 1:24 scale NASCAR dispenser. These are pictured later in the book and have minimal collector value at the moment due to their widespread continued availability.

Packaging for the large dispensers varies throughout the world. This one originated in the Far East.

-10" - International

Sax player - Peanut Green		$40-60
Sax player - Peanut Yellow		$40-60
Sax player - Peanut Red		$40-60

'-10" - International

Dispenses into hand - Plain Red (new face)		$20-40

M&M's Racing Team 1:24 scale plastic race car dispenser, available at retail outlets in 1999. $8-12

45

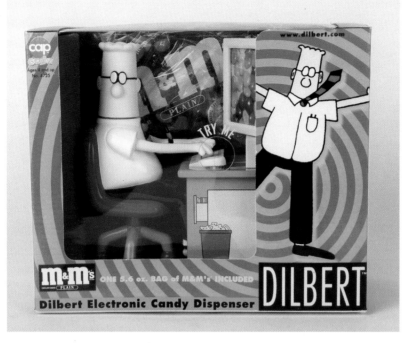

This great Dilbert Electronic Candy Dispenser was available at retail outlets during the 1998 holiday season. $15-25

Early metal and glass M&M's coin operated dispenser. Later version were made of plastic. Rare

This 9" plastic dispenser from 1991 was available at retail stores. In the late 1990s, it was modified to include a colorful M&M character on either side of the "M&M's" logo. Each $5-15

"Enjoy the Great Taste" slogan is on every side of this swivel-based Snack Center. Offerings include Snickers, M&M's Plain and Peanut Chocolate Candies! Rare

Two European 7" dispensers on original card packaging. $30-35

1999 10" tall plastic New Millennium Dispenser (shown in catalog photo) lights up and can dispense up to three different types of candies, depending on which character's arm is pulled. This dispenser was available to Inner Circle™ members through their catalog. $20-30

1999 Orange Hide 'N Hander dispenser appeared late in the year. $8-12

This Disney battery operated candy dispenser was only available at Walt Disney Theme Park Gift Shops. $25-40

This POWER battery operated dispenser was sold at retail stores in three colors: red, blue, and yellow. $8-15

The Millennium battery operated dispenser was a Limited Edition dispenser celebrating the "MM means 2000" promotion. $10-20

These "pull down the hand" type plastic candy dispensers hold full size candies. Called "Fun Size Candy Handlers," they were available in retail outlets during 1999 and are similar in operation to the mini-dispensers. $8-12

Displays

Displays are the advertising appeal that catches the candy cravings of all consumers. They come in all sizes, shapes and types of material.

Six foot tall store display is eye appealing in yellow! Rare

This 26 1/2" tall Snickers display rack from 1997 has metal shelves and plastic sides to hold boxes of various Mars candy products, including M&M's. $75-100

Large, chest high, Blue Candy standing store display from 1997 has plastic shell. Original versions had Styrofoam arms and legs; later versions had plastic shell arms and legs as well. Each $350-700

Two inflatable, vinyl figures.
Various colors have been
available in 26"-30" heights.
Each $35-55

Over 3' tall, this battery operated, animated store
display features a wobbly tree decorator. Rare

Displays - Danglers

Display danglers are the point of purchase eye catchers that lead consumers to the end product! They are appealing little 6"-8" clear plastic flaps that hang off the shelves in minute markets, gas stations, retail stores, grocery stores, and everywhere M&M/Mars products are sold. Each $3-8

51

Displays - Two-Sided

Blue and Red lead the way on this two-sided, molded plastic, advertising display piece! It originally sat on top of a stick and encouraged everyone to try the latest candy selections! $30-50

Yellow (peanut) is all decorated for the M&M's Easter advertising season! This 2' tall, two sided, molded plastic display is just the right M&M Bunny to attract consumers. $30-60

Yellow is King of the M&M's product line in this two sided, molded plastic advertising display. Flag selections include: Snickers, Skittles, M&M's, M&M's Peanut, Milky Way, and Starburst. The three foot tall display soars above the counter perched on the stick attached to one foot. $40-60

This two sided, molded plastic advertising display has Red and Green atop a stick, greeting shoppers. $30-50

52

Displays - European Boxes

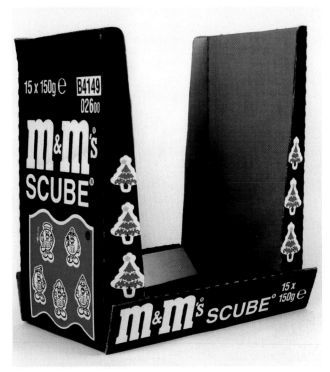

A nice collection of rare M&M's flag toppers are stored in this M&M's plain and M&M's peanut box from the United Kingdom!

M&M's Scube tubes (the European name for tubes with toppers) came in this holiday display box.

Dr. Pepper

An 8-ounce Dr. Pepper bottle was filled with M&M's (or Jelly Belly candy) and sealed at the bottling plant for a 1998 MARS Christmas party. It has a sticker on the bottom showing the tie-in with M&M's & Jelly Belly Candies. $10-15

Egg Cups

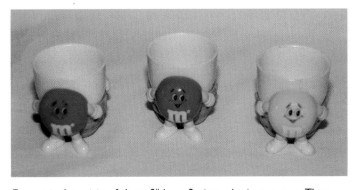

Europe is the origin of these 3" long, 2-piece plastic egg cups. They come in red, green, and yellow. Each $20-35

Errors or Fakes

Can you help the authors solve these mysteries?

Above: This photo of the small European brown soccer player (peanut) with white colored arms and with flesh colored arms leads one to wonder if both were manufactured in quantity.

Left: This American dispenser with yellow colored hands and feet is an unusual item! Could its appearance be due to sun fade, paint, or smoke?

This orange 7" soccer player is an interesting item since the soccer player was not available commercially in orange. Perhaps the foot peg holding a present was replaced with the soccer ball?

Fabric

Many different patterns have been available at fabric stores showing the M&M characters and/or candies. Over twenty are known to exist. This picture shows two of the choices available for the 1999 Christmas season. Per yard $4-8

First Day Covers

Two 1997 First Day Covers by H & H celebrate the holiday season! Less than fifty of each are reported to have been made. Each $15-25

FTD

Valentines Day 1999 would not have been the sweetest without this ceramic FTD Floral container! $15-25

Any occasion would be perfect for this yellow FTD 1999 ceramic container. $10-20

his 3" pinback FTD button says it all: "Ask Me About the FTD weet Surprise Bouquet." $5-15

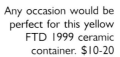

merican FTD florists used these red and yellow plastic containers and matching figures ith gift card for a surprise bouquet offering. The Green figure was available in Europe, ut without a matching pot. See dispenser listing for pricing of figures by themselves. ed or Yellow Set $15-30

This 1999 FTD ceramic basket is perfect for gathering flowers and M&M's candies left by the Easter Bunny! $10-20

The 1998 holiday season saw Red and Yellow helping Santa bring cheer down the chimney on this ceramic offering. Shown with Valentines Day heart container from previous page. $15-25

Halloween

1999 Light-Up Pumpkins are made of hard Styrofoam, range up to 12" tall, and are available in orange, red (not pictured), yellow and blue. $8-12

1999 plastic 12" Boo Beams are actually glowing Halloween wands. Batteries go in the handle and kids carry them trick-or-treating. They are available in yellow (as a mummy), red (as a devil), green (as a witch), and orange (as a pirate). $5-10

1999 Trick-or-Treat plastic pails are up to 7" tall and perfect for youngsters to use for gathering Halloween treats. They are available in yellow (as a mummy), red (as a devil), green (as a witch) and orange (as a pirate). $5-10

1999 4 1/2" plastic Frightful Lightfuls can be clipped on or worn as a necklace. They light the way for trick-or-treating. They are available in yellow (as a mummy), red (as a devil), green (as a witch), and orange (as a pirate). $5-10

Hats

999 Mini Candelabra figures are 6" tall nd each has two suction cups so it can e stuck on a window. The figures ome with a flickering candle styled ght and cord and are available in green the witch) and red (the devil). $10-15

These hats were available with Blue, Green, Red, and Yellow on the front; the bill color matches the color of the figure. Each came prepackaged with M&M's candies for additional enjoyment. $8-12

One of the earliest advertising programs began with this Styrofoam hat reading "Vote For Fun in America!" The advertising campaign included hat, button, and banner. Set $30-60

M&M/Mars produced a variety of hats in assorted designs and illustrating various characters over the years. Here is a sampling of some of them.

Jewelry

This "M" Pendant, in 14K solid gold with a gold chain, was available in the Employees catalog and the Inner Circle™ catalog from Mars. $100-150

This pendant is a green "M" with gold tone chain. $25-50

Key Chains

M&M/Mars produces an assortment of key chains around the world. This selections represents a sampling of some of the creative ones produced.

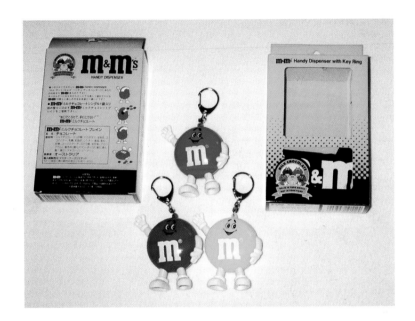

M&M's World, Las Vegas produced an assortment of 2" key chains representing each character. $4-8

These 1998 plastic European Funny Lights are a little over 2" tall and come on a blister pack. $20-35

Above: This Blue key chain has a built-in pocket light. $15-25

Right: This Flip Flashlight in Blue from 1997 was a fundraising item from Europe. $15-30

1999 Japanese produced plastic keychains stand 2" tall. $10-20

Kitchen & Household

Salt and pepper shakers, circa 1997. $25-40

"Bake with the Best" Pot Holder is from 1986. Note the early characters. $15-30

Yellow and Orange early kitchen magnets. Note the skinny styling in the "M&M's" logo, indicating a pre-1990s product. Rare

"Leader of the Packs" rubber welcome mat. $25-35

These 8" ice cream spoons from M&M's World are topped off with a small PVC character. Each $5-10

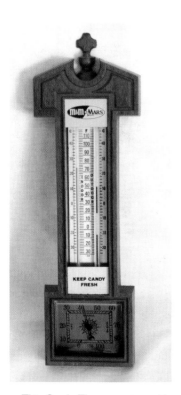

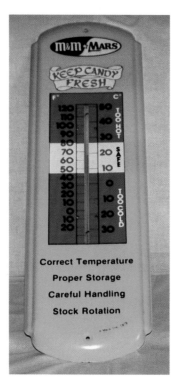

This Candy Thermometer, with centigrade and indoor/outdoor readings, is apparently a later model than the one at right. Rare

1979 M&M/Mars "Keep Candy Fresh" Candy Thermometer was an early attempt by Mars to encourage (as noted) proper storage of its product, proper temperature control, careful handling, and stock rotation for customer satisfaction! Rare

A 3" plastic cookie cutter was available in a couple of the tins. $5-10

Lights - Holiday

1993 Set of Christmas Happy Lights without the Blue figure (not yet introduced). $25-35

1996 Set of Christmas Happy Lights with the Blue figure, plus set of orange Halloween Lights. Each $20-30

Above and right: 1998 Set of ten Happy Lights with only five lights showing in the packaging. M&M's Red advertises that "Most Stars only get their name in lights!" $15-25. Second set is a 1998 Happy Lights Set with all ten lights showing on the packaging, the rarer packaging of the two. $20-30

1999 Blue packaging for the same light set introduced in 1998. $12-15

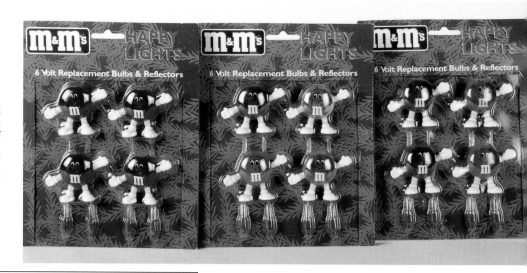

Replacement Lights sets showing all three variations of colors offered: four Blue; one each of Red, Green, Yellow, and Orange; and Holiday pack with two Red and two Green. Each $5-15

Luggage Tag

4" Plastic luggage tag with Yellow on the move. $15-20

Lunch Box Items

These 3", liquid filled, lunch box coolers from 1997 come in four colors: Green, Blue, Red, and Orange. They freeze and keep food and candy at just the right temperature and are advertised as FUN ICE! Each $4-8

These 7 1/2" tall, 7" long, plastic, yellow lunch boxes from 1997 come with a green or blue thermos top. Made in Europe, they hav[e] a nylon carrying strap attached. Each $25-45

64

Minis

Minis were introduced in 1996 and are tiny M&M's sold in a tube container. The characters are called Swarmees and were sold as plush ~~ures~~.

~~99~~ Minis tubes came packaged with a 3 3/4" x 5 1/2" cardboard ~~orybook~~. Each $3-6

There were two different sets of these dispensers, with the newer light blue card set being a little harder to find. The three colors of the earlier dark blue card "Candy Handers" are Pink, Green, and Blue (pictured), while the later (light blue card) set was available in Orange, Yellow, Blue, and Red. $8-12. In late 1999, larger Candy Handers were introduced in the Yellow and Red characters for the regular sized M&M's candies. Each $10-20

1999 "RIGS" were plastic trucks that dispensed Minis from tubes on top of the truck. They were available in Red, Green, Blue, and Yellow from retail outlets. Each $4-6

Minis Counter Display Dispenser. Note the FREE HAT mail-in offer for a black children's size hat with sewn-on cloth patch.

1998 saw the arrival of the first generation of plush Swarmees at the retail chains. These 3" wide figures are the characters from the Mini-M&M's and have Velcro hands for the soft plush collectibles "to stick together," forming a chain in Red, Purple, Blue, Yellow, Orange, Green, Pink, and Brown. Pictured is one complete set of the first generation in the display box. Each $5-10. A second generation of Swarmees was introduced in the fall of 1999 (not pictured).

An interesting set of four Mutant Mini Swarmees appeared in the summer of 1999. Each of the fou characters offered (Luminous Larry, Neon Newman, Swirlin' Merlin, Wavy Daisy) has a differe facial expression. The color variations of each character are endless and no two of the same character are alike in terms of th material colors. Each has a tush t with the character's name. Each $6-12

Mini Figures

Mini figures (1" tall) from Europe came with stick-on decals with assorted facial expressions. Each $5-10

Other small figures (under 1 1/2" tall) have been found. They are of unknown origin; and there are probably more than the ones pictured. Each $3-8

Mini Figures - Button Covers

These European plastic mini figures, under 1" tall, are clothing button covers. They add just the right touch of FUN to a shirt or blouse, in four colors. Set $20-35

Mugs

Many mugs, in various colors, sizes and shapes have been offered. Following is an assortment of what the authors believe to be some of the cuter ones! Various packaging highlights the mug selection, from mugs with Swarmees to mugs bundled with M&M's plain and peanut packages.

M&M's Brand mugs and the round base collection originate from M&M's World in Las Vegas. Mugs came packaged with a Bendy or plush and candy packages during the 1998 holiday season at two large retail chains with exclusive licensing arrangements. Various national and international retail stores produce an ongoing array of M&M's collectibles.

68

Olympic Items

1992 Olympic metal key chain and pin.
Each $10-20

M&M/Mars presented these ribbon adorned plaques to participants who "Helped America Win...U.S. Olympic Team." Rare

The 1984 Olympics accelerated the parade of sports-related merchandise from M&M/Mars: two different glass Olympic Jars were produced. The smaller of the two (7 3/4" including plastic rimmed stopper) has the two figures with one flag. $10-25. The taller (8 1/2" including the stopper) has the same two figures, but each is holding a flag on this one. The larger one also announces that they are a "Sponsor of the 1984 Olympic Games." $15-30

1992 Olympic hat and fanny pack marked "Worldwide Sponsor 1992 Olympic Games." Hat $10-15; Fanny pack $15-30

1992 Summer Games Commemorative Pin Collection, twenty-six different pins in a velvet covered 7 1/2" x 7 1/2" case. Set $50-75

70

1992 Olympic ceramic mugs and plastic watch.

992 Winter Games Commemorative Pin collection, ten different pins in a velvet overed case. Set $25-40

Packaging

Packaging is not the average collector's item of choice; "Interesting but not always collectible" is the common phrase heard on the collecting trail! Nevertheless, pictured are the authors' selection of some of the more collectible packaging items.

These are a selection of some of the many European produced Candy Action Boxes.

Cardboard vehicles decorated as a truck, school bus, sled and Easter Bunny each contain ten small boxes of M&M's. They were offered over the course of a year or so for various holidays or times of the year, circa 1996. They measure 6" x 3" x 2 1/2". Each $15-25

Specially designed 1 3/4" x 3 3/4" box for distribution only at the White House and Air Force 1 during President Bill Clinton's tenure. $15-30

Specially designed 1 3/4" x 3 3/4" box advertising Miss Hall's School. Rare.

Cardboard "Happy Valentine's Day" box from Europe, 6" wide. Rare

1992 German Easter Holiday Egg made of papier mâché like material. $15-25

72

English (UK) Easter Holiday candy egg in package. Chocolate egg is gold foil. Also included are packages of M&M's plain and peanut candy. $10-15

This papier mâché, heart-shaped container held candy and toys from the early '90s. $15-25

This 1998 European Mini Mix Box (7 1/2" x 3 1/2" x 3 1/2") held a selection of M&M's Chocolate and M&M's Peanuts. $5-15

Ronald McDonald and M&M's teamed up in France for this promotional package. $5-10

Right and next page: 1990s Packaging - around the world!

Party Supplies

Pictured is a variety of party supplies that are too numerous to list individually. Many are still available at retail outlets.

78

Above: This 10 1/2" ceramic, 2-sectioned party snack bowl has a great picture of two characters on a football field on one side of the bowl. $25-50

Left: This 1999 set of four Straw Huggers was sold at M&M's World, Las Vegas. Set $20-30

Pins

This set of nineteen pins pictured the flags from countries participating in the World Cup Soccer Tournament. Rare.

This "Grab On To That M&M's Feeling" pin was part of a "Vote For Fun In America" promotion that included a hat, button, and banner (see page 58).

Assortment of M&M's Brand pins.

Plush - Stuffed Characters

Many sets of plush toys have been offered over the years in all shapes, sizes, materials, and facial expressions. Here is our understanding of the groupings. Generally speaking, older and/or bigger plushes command a higher value than smaller and/or newer ones.

Ace, 6" body with dark blue shoes and blue tag.

Caltoy Plain and Peanut, 1987.

Ace, 7" body with dark blue shoes and black tag.

Ace, 6" body with light blue shoes and black tag.

Mary Meyers Fun Friends,
8", Plain, 1994.

Mary Meyers Fun Friends, 10",
Peanut, 1994.

Ace, 4 1/2" body, black tag
with "M" screen printed.

Ace, 5" body, black tag.

Fun Friends Plain, 3", 1994.

Fun Friends Peanut, 3", 1994.

3" Red Santa, Green Santa, Blue Baker, and Lilac Bunny.

3" Golfer, 1997 Witch, 1996 Witch, and Tennis Player.

Finger Puppets, plain and peanut —
complete set of eight.

Bean Bag round plush,
M&M's World, Las Vegas.

82

ed Plain, 24" — available in other colors also.

Large plush yellow peanut with blue sneaker accents.

ean bag plush from M&M's World, Las Vegas (green not pictured).

Large plush peanut with white sneakers, no accents.

Holiday Mini Plush figures have been offered for various holidays through mail-in promotions and retail store sales. This Christmas set all have Santa hats. Similar sized plushes were available in Easter and Halloween accessories. A set was also issued with a golf theme.

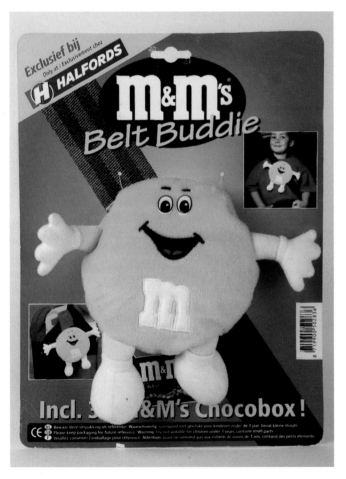

These four 1999 Talking Plushes each say two different phrases when you squeeze their stomachs.

This Belt Buddie from Holland fastens to the seatbelt and has a zippered pouch for valuables. $30-50

Pocket Surprises (Pochette Surprises)

These interesting little M&M toys were initially offered in France in conjunction with the International Toy Exposition, held in Paris a few years ago. A second and third series were subsequently issued. Pocket Surprises were a variety of toys packaged in a cellophane bag along with instructions on how to put together or use the toy, a colorful insert showing other toys available in that set, and peel-off decals if appropriate to the toy. This cellophane bag was packaged in a 2-piece plastic case, and then both were packaged in a foil bag, along with 15-20 M&M's.

The sets, consisting of single-piece or 3-D put-together figures, were usually available in three different color variations, while items like the dexterity puzzles, flat, put-together figures (soccer and circus), glow-in-the dark set, carnival sets, and others have no color variations. The three series offer about 140 different toys, with the total rising to about 330 when all the variations are included. Many collectors are only interested in finding one of each toy, while other collectors want all the variations as well.

Pocket Surprises Series I, mint-in-package.

Pocket Surprises Series II, mint-in-package.

Pocket Surprises Series II counter card from France.

Pocket Surprises Series III packaging.

Soccer - One-Piece Figures
(all have three color variations except puzzle)

❏	Goalie	$3-5
❏	Fallen Down	$3-5
❏	Cup Holder	$3-5
❏	Kicker	$3-5
❏	Fan	$3-5
❏	Yellow Card	$3-5
❏	Jigsaw Puzzle to Match Insert is Toy from Series II	$5-8

Soccer - Put Together

❏	Goalie	$3-5
❏	Kicker	$3-5
❏	Runner	$3-5
❏	Referee	$3-5
❏	Fan	$3-5
❏	Header	$3-5

Medieval (all have three color variations except puzzle)

- ☐ King with Goblet & Building with Cross Shaped Windows $3-5
- ☐ Joker with Building with Slit Windows $3-5
- ☐ Guard with Long Ax $3-5
- ☐ Knight on Horse $3-5
- ☐ Guard with Shield and Sword $3-5
- ☐ Jigsaw Puzzle to Match Insert is Toy from Series II $5-8

Circus - Solid Figures
(all have three color variations except puzzle)

- ☐ Clown with Saxophone $10-20
- ☐ Clown with Fiddle $10-20
- ☐ Magician with Rabbit $3-5
- ☐ Strong Man $3-5
- ☐ Elephant Trainer $3-5
- ☐ Elephant $3-5
- ☐ Horse $3-5
- ☐ Lion Tamer $3-5
- ☐ Lion $3-5
- ☐ Jigsaw Puzzle to Match Insert is Toy from Series II $5-8

Circus - Put Together
- ❏ Three Acrobats $3-5
- ❏ Magician $3-5
- ❏ Drummer $3-5
- ❏ Juggler $3-5
- ❏ Ring Master $3-5

Space (all have three color variations except puzzle)
- ❏ Flag Holder $3-5
- ❏ Dish Holder $3-5
- ❏ Walkie-Talkie Holder with Six-Wheeled Vehicle $3-5
- ❏ Thumbs Up With Two-Wheeled Vehicle $3-5
- ❏ Jigsaw Puzzle to Match Insert is Toy from Series II $5-8

Prehistoric Vehicles & Dinosaurs
(all have three color variations except puzzle)
- ❏ Rider with Jacket & Standing Dinosaur $3-5
- ❏ Dinosaur with Caveman $3-5
- ❏ Caveman with Log Car $3-5
- ❏ Caveman with Flatbed Car $3-5
- ❏ Jigsaw Puzzle to Match Insert is Toy from Series II $5-8

City Playsets (all have three color variations)
- ❏ Skating $3-5
- ❏ Jammin' $3-5
- ❏ Hot Dog Vendor $3-5
- ❏ Soccer $3-5
- ❏ Basketball $3-5
- ❏ Skateboarding $3-5

Cowboys & Indians Playsets (all have three color variations)

- ❏ Indian Village $3-5
- ❏ Indian Camp $3-5
- ❏ Arrest at the Restaurant $3-5
- ❏ Stagecoach Holdup $3-5
- ❏ Robbery at the Hotel $3-5

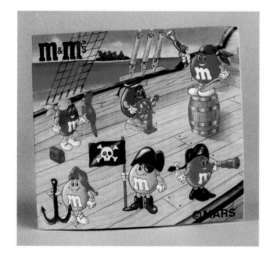

Pirates (all have three color variations except puzzle)

- ❏ Pirate with Parrot $3-5
- ❏ Pilot with Ship's Wheel $3-5
- ❏ Pirate with Musket and Sword $3-5
- ❏ Pirate with Anchor $3-5
- ❏ Pirate with Flag $3-5
- ❏ Pirate with Eyeglass $3-5
- ❏ Jigsaw Puzzle to Match Insert is Toy from Series II $5-8

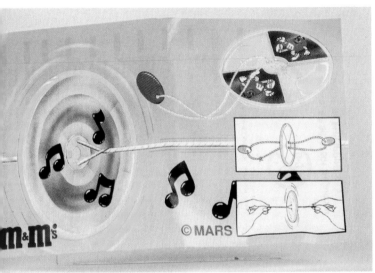

Miscellaneous Sets Not Related

- ❏ Top — $3-5
- ❏ Spinning Disk — $3-5
- ❏ Tiddly Winks — $3-5

Dexterity Puzzles - I

- ❏ Basketball $3-5
- ❏ Bear $3-5
- ❏ Dragon $3-5
- ❏ Elephant $3-5

Dexterity Puzzles - II

- ❏ Haunted House $3-5
- ❏ Treasure Cove $3-5
- ❏ Bottom of the Ocean $3-5
- ❏ Outer Space $3-5

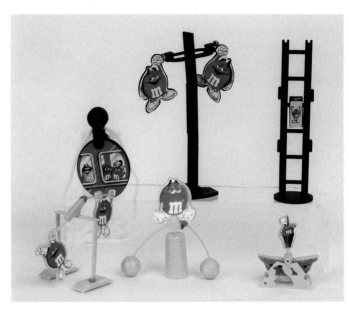

Balance Toys

- [ ] Cable Car — $3-5
- [ ] Two on Click Down Ladder — $3-5
- [ ] Ladder Gravity — $3-5
- [ ] Two On Balance Bar — $3-5
- [ ] Cone Balance — $3-5
- [ ] Pirate Ship — $3-5

Carnival Rides

- [ ] Ferris Wheel — $3-5
- [ ] Merry-Go-Round — $3-5
- [ ] Space Sensation — $3-5
- [ ] Swinging Basket — $3-5
- [ ] Balancing on High Wire — $3-5
- [ ] Strongman — $3-5

Caveman (all have three color variations)

- [ ] Cook with Large Drumstick — $3-5
- [ ] #1 — $3-5
- [ ] Head Scratcher — $3-5
- [ ] Trike Rider — $3-5
- [ ] Sculptor — $3-5

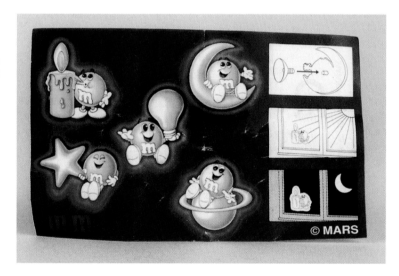

Glow-In-the-Dark

- [ ] With Candle — $3-5
- [ ] Sitting on Crescent Moon — $3-5
- [ ] With Light Bulb — $3-5
- [ ] With Star — $3-5
- [ ] Sitting on Saturn — $3-5

Occupations (all have three color variations)
- Artist with Easel — $3-5
- Chef with Tray — $3-5
- Doctor with Needle and Bag — $3-5
- Reporter with Mike and Tape Recorder — $3-5
- Detective with Magnifying Glass and Pipe — $3-5
- Fireman with Extinguisher and Fire — $3-5

Winter Sports (all have three color variations)
- Sledder — $3-5
- Snowball Thrower — $3-5
- Skier with Skis — $3-5
- Snow Boarder — $3-5
- Hockey Player with Stick — $3-5

Road Signs I (four packs, two per pack - six total, some doubles)

❏	Train	$3-5
❏	Stop	$3-5
❏	Falling Rocks	$3-5
❏	Haunted Castle	$3-5
❏	3.50 m	$3-5
❏	Traffic Light	$3-5

Road Signs II (four packs, two per pack - six total, some doubles)

❏	Bumps	$3-5
❏	Food	$3-5
❏	Paris	$3-5
❏	Trophy	$3-5
❏	Airport	$3-5
❏	Do Not Enter	$3-5

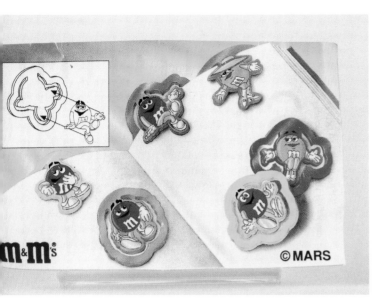

Book Clips (five packs, three per pack - fifteen total)
- ❏   Pack 1 $3-5
- ❏   Pack 2 $3-5
- ❏   Pack 3 $3-5
- ❏   Pack 4 $3-5
- ❏   Pack 5 $3-5

Advertising for miscellaneous
sets shown on the next page

Miscellaneous Sets Not Related

- ❏ Stencil Set of 4 $3-5
- ❏ Hockey Player Stencil $3-5
- ❏ Paint Palate $3-5

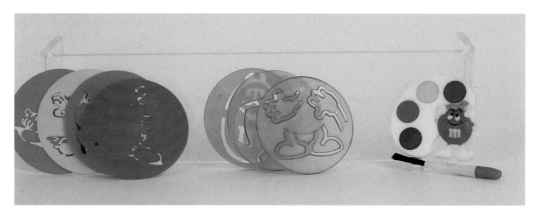

Band (all have five color variations)

- ❏ Drummer $3-!
- ❏ Xylophone $3-!
- ❏ Trumpet $3-!
- ❏ Violin (not pictured) $3-
- ❏ Saxophone $3-
- ❏ Conductor $3-!
- ❏ Flute Player $3-!
- ❏ Cello Player $3-

Beach Scene (all have five color variations)

❏	Surfer	$3-5
❏	Sailor in Raft	$3-5
❏	Sailboarder (not pictured)	$3-5
❏	Raft with Drink	$3-5
❏	Innertube	$3-5
❏	Beach Chair (not pictured)	$3-5

Postcards

Postcards by Mars continue to be produced around the world. Pictured here is a sampling.

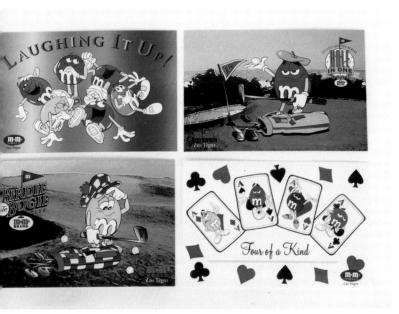

Four postcards from M&M's World, Las Vegas. Each $4-8

Two postcards from the USA. Each $5-10

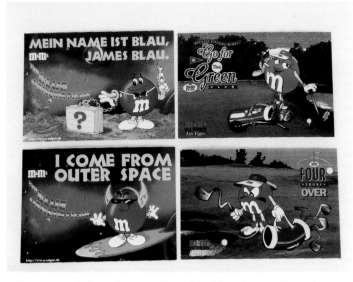

Two postcards from Germany (Each $10-20) and two additional ones from M&M's World, Las Vegas. Each $4-8

Two postcards from Europe. Each $10-20

Prints

M&M/Mars produced a series of three prints: Blue, Green, and Red/Yellow. Each $60-120

Racing Team Items

In 1998, the #36 Skittles car, driven by Ernie Irvan, was repainted for the Las Vegas Race as an M&M's car. Racing Champions, Revell, Action, and other licensed manufacturers scurried to get a few M&M's cars and memorabilia out to meet the demand. Generally speaking, this means that the 1998 cars (the #36 logo is painted on in green paint) should be harder to find, and therefore more highly valued than cars produced in later years. In 1999 (the #36 logo is painted on in blue paint) the M&M's car was one of the circuit regulars, and the 1999 cars and memorabilia were much more plentiful. The 1999 car was repainted for one race as a blue Crispy car, and again as the Pedigree car, making these two issues more difficult to find in 1999 (identified by the blue #36). In September of 1999, Ernie Irvan announced his retirement. Only time will tell if that makes all of his M&M memorabilia worth more.

Pictured is an assortment of racing collectibles, some of which were available in NASCAR outlets and others only through mail-in offers, membership in the Ernie Irvan fan club, or at the individual NASCAR races. The values of the die-cast cars vary greatly. They seem to carry prices from retail to three times retail, depending on where you find them. Because they are a new item, no further attempt at pricing has been attempted for this edition.

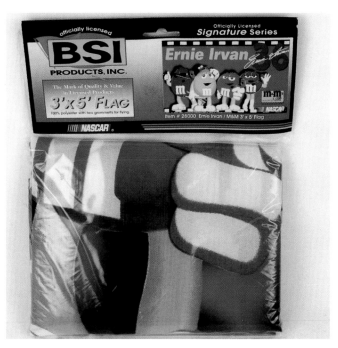

1999 Racing Team Flag, available at NASCAR outlet stores. $25-30

'98 Racing Team kid's jacket. $75-150

Racing Team bumper sticker and 1999 Press Pass Racing Cards set, available at retail outlets. $4-8

1998 Bumper sticker, available at NASCAR outlets. $5-10

Ernie Irvan key chains. $6-12

1999 M&M's Racing Team trading cards. $3-6

1999 Racing Team seat cushion (shown front and back), available at NASCAR races only.

1999 Racing Team rain parka, key chain and child's bib, only available at NASCAR races

Large 1999 vinyl window poster. $10-20

1999 Bumper sticker. $4-8

1999 Ernie Irvan magnets. $6-12

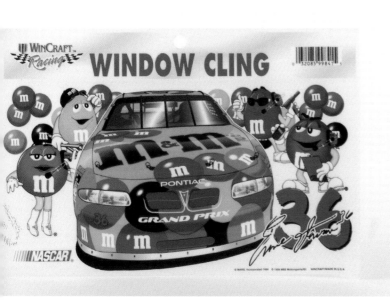

999 Racing Team window cling. $15-30

Top: 1998 license plate. Bottom: 1999 set of two license plates. Each $8-15

1999 Crispy license plate. $6-12

1999 NASCAR, Pin #36. $6-12

Elite, 1998, 1:24 scale, yellow #36
NASCAR car with gold coin.

Action, 1998, 1:32 scale, Racing Collectables Club of America (1 of 700) boxed gold car.

Action, 1998, 1:64 scale, yellow blister pack model stock car #36

Racing Champions 50th Anniversary Editions, 1998: 1:64 scale, set of Pontiac Grand Prix gold chrome and yellow cars with certificate included in the boxed set (1 of 5,000).

Racing Champions 50th Anniversary Editions, 1998. (L to R): yellow transporter with M&M's yellow car on tailgate. The size of this transporter seems to be between the standard 1:64 scale and 1:44 scale, with size not denoted on the packaging; 1:64 scale, yellow transporter boxed car; 1:144 scale, yellow transporter with yellow car on top of trailer in blister pack.

Left: Racing Champions 50th Anniversary Editions, 1998. Top Row (L to R): 1:144 scale, set of five cars with M&M trading cards showing in blister pack; 1:64 scale, yellow car with racing card behind auto in blister pack. Bottom Row (L to R): 1:64 scale, yellow Stock Rods car with collector's card behind car (1 of 19,998) in blister pack; 1:24 scale, yellow die cast stock car replica boxed car; 1:144 scale, yellow Stock Rods car with collector's card in blister pack.

Below: Racing Champions 50th Anniversary Edition, 1998. Close-up of 1:144 scale, set of five cars. Note M&M's cars in lower left corner.

Revell, 1998. Shown is an assortment of Revell's regular and select yellow cars, ranging in size from 1:64 to 1:24 scale, boxed. Note: 1:24 scale size is 1 of 5,998.

Above and Right: Racing Champions 50th Anniversary Editions, 1998: 1:24 scale, set of gold chrome and yellow cars with certificate included. Packaged in two separate boxes; when joined they form one scene.

Racing Champions 50th Anniversary Editions, 1998: Left: 1:24 24K gold car boxed. Center: 1:24 scale gold boxed bank; 1:24 scale, gold boxed car; 1:64 scale, 24K gold car in blister pack. Right: 1:64 scale, Special Edition Toys R Us retail chain (1 of 9,999) gold car in blister pack.

1999 Hot Wheels, 1:24 scale, blue Crispy car with 1:64 scale yellow #36 M&M's car, boxed set.

1999 Action, 1:24 scale, Ernie Irvan #36 M&M's yellow car, yellow boxed.

1999 Action, 1:64 scale, Ernie Irvan #36 M&M's yellow car, blue boxed.

1999 Revell, 1:64 scale, Pedigree car.

1999 Action, 1:64 scale, Crew Chief, Jack Man, Business Manager, and/or Tire Changer with #36 Ernie Irvan's yellow car, blister pack. Cars are identical, mounting cards are different characters.

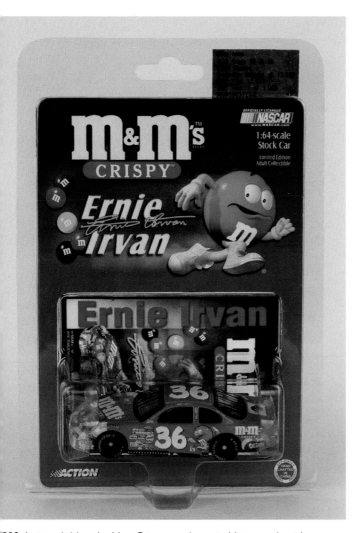

1999 Hot Wheels, 1:64 scale, Pro Racing #36 yellow M&M's car in blister pack.

1999 Action, 1:64 scale, blue Crispy stock car in blister pack with collector's card.

1999 Hot Wheels, 1:64 scale, Track Edition, NASCAR, yellow #36 M&M's boxed car.

1999 Hot Wheels, 1:64 scale and 1:24 scale, #36 yellow M82 Motorsports, Pontiac Grand Prix boxed cars.

1999 Racing Champions, 1:64 scale, Commemorative Series gold chrome (1 of 9,999) gold #36 car, in blister pack.

1999 Racing Champions, 1:64 scale, Press Pass (1 of 19,999), yellow stock car #36, in blister pack.

1999 Racing Champions, 1:64 scale, The Originals yellow #36 car packaged in tw different background blister packs (helmet placement variations).

1999 Racing Champions, 1:64 scale, silver chrome Special Edition, Toys R Us (1 of 9,999) silver car #36, in blister pack.

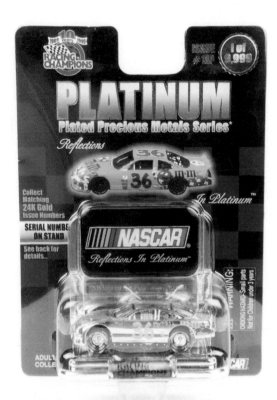

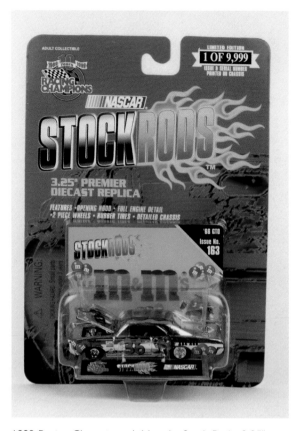

1999 Racing Champions, 1:64 scale, Reflections in Platinum, Platinum Plated Precious Metals Series (1 of 9,999), platinum car, in blister pack.

1999 Racing Champions, 1:64 scale, Stock Rods, 3.25" Premier Diecast Replica, gold car, in blister pack.

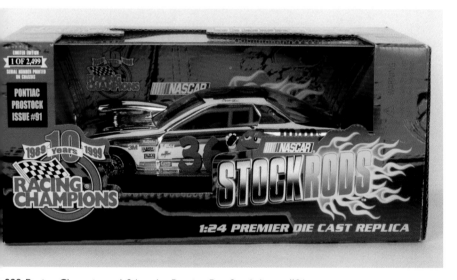

999 Racing Champions, 1:24 scale, Pontiac Pro Stock Issue #91 Stock Rods gold (1 of 2,499), boxed car.

1999 Revell, 1:64 scale, yellow #36 (1 of 5,000) boxed car.

1999 M&M's truck with trailer and yellow #36 car, available from QVC. Later in the year, a similar set featuring the blue Crispy car was also available from QVC (not pictured).

1999 Racing Champions, 1:64 scale, transporter with yellow, gold and silver 1:64 scale cars, boxed set.

Ernie Irvan #36 issued his own "Prepaid Phone Card." M&M's also offers phone cards with each character pictured.

Radios

1998 European 3 1/2" "Winking" Radio. $30-50

Above: 1999 10" plastic, motion activated AM/FM blue character radio. $40-75

Right: 1998 European boom box FUN Radio. $35-60

997 1" Blue mini radio with ear piece. $15-25

This 2 1/2" tall Christmas ornament (top) and set of three 2" magnets were designed as school fundraiser premiums for the 1996 holidays in USA schools. All were made of polyresin. Ornament $10-15; Magnets $15-20 each

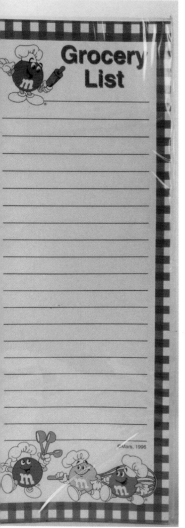

996 Grocery List pad is 11" x 4 1/4". $10-15

1996 Strip of flat magnets. $10-15

The 1996 "Bright Ideas" and 1997 "Sweet Surprises" school fundraiser catalogs and order forms are eight pages and six pages long respectively and measure 11" x 8".

1996 Ceramic mug and utensil holder, tin recipe box, and coupon holder; all designed for fundraising activities through schools in the USA. Each $10-15

1996 set of three rectangular tin magnets, 1 3/4" x 2 1/2". Each $5-10

Hot pad ($10-15), hand towel ($10-15), Things to Remember list ($8-15), tin (available in several colors, $5-10 each), and mug ($10-15) were all 1996 school fundraiser items in USA schools.

Heavy cardboard "Messages" board from 1996 measures 9" x 12" and comes with a black marker pen. $15-20

1997 School bus tin opened in the back to hold candy. $10-20

Five foot Watch Grow chart was another 1996 fundraiser premium in USA schools. $15-20

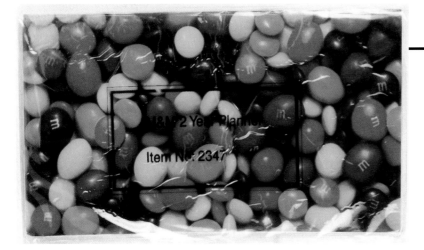

School Supplies

M&M's Two year pocket planner.

DS Mouse Pad. Other round, 8 1/4" wide mouse pads have been produced in red (nd possibly other colors) with the "M" logo in the middle. $20-30

Lucite picture frame, 3 1/2" x 5". $15-20

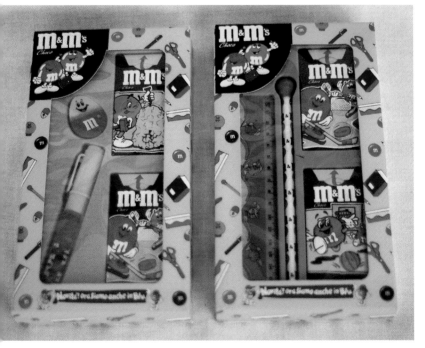

996 European school supplies kits, with pen and/or pencils and ruler plus two ction boxes of M&M's Chocolate Boxed Candies. $30-45

1998 Crispy notepad is 4" x 6"— "The Feeding Frenzy Has Begun!" $6-12

Kid's zippered 8" nylon Smile Sack — designed to hold school supplies or other kids' collectibles! $15-20

Mars has produced many pens and pencils. Here are some of them.

"M&Ms"® Brand Character Pen

ENSEMBLE STYLO

Shopping Bags

Shown is an M&M's World, Las Vegas shopping bag plus an assortment of other USA shopping bags and an Australian green Christmas bag.

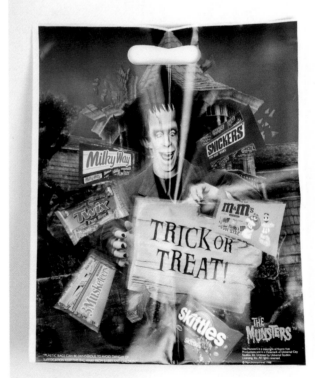

Shot Glass

1999 M&M's World Las Vegas commemorative gold colored glass shot glass with the M2 logo on the front. $8-15

Stickers

These individual 3" paper stickers came from Europe. They were issued to introduce "Blue" to the world. Each $3-5

MEIN NAME IST BLAU JAMES BLAU
m&m's ®

I COME FROM OUTER SPACE!
m&m's ®

LIEBER BLAU ALS BLOND!
m&m's ®

ICH BIN BLAUMANN!
m&m's ®

Set of 1992 kids stickers plus another sheet of stickers dated 1995.

1998 Fun Deal set of 18 stickers was available in retail outlets at Christmas. $2-5

Late '80s to early '90s French set of 30+ stickers came packaged 5-6 per pack. When peeled off and applied to a promotional Sticker Book (not shown), they told a story. Each sticker is individually numbered. Pack $5-10

Sun Visors

1986 Accordion-style
M&M's sun visor. $25-50

1996 Accordion-style
M&M's sun visor. $20-30

Telephones

1999 Hands-free, plastic Red Plain guy and Green Plain lady figures can be hung around your neck. Put on the earphone and you are ready to walk and talk as far as the 18-foot cord will allow. Sold in blister packs. Each $25-40

1999 Time Capsule Kit is a two-piece gold plastic capsule filled with M&M's candies and other items. $20-30. An earlier, silver colored capsule is believed to exist, and would be considered quite rare.

Time Capsule

The Time Capsule contains one yo-yo, one hi-bounce ball, 2000 stamper, artifacts envelope, likes and dislikes list, hints book, ID tags and twelve stickers.

Tins

Tins have been issued during the holiday season each year since 1986. In addition, special tins have been issued as fundraisers for other holidays or various other reasons. Most of the holiday tins are fairly common and have a value of $3-10. Some of the other tins (Save the Children tins, NATCO tins, cardboard containers, etc.) may bring a much higher price.

1986 Round Yesteryear Town Scene tins, plain and peanut, 13 and 24 oz. (two with plain brown lid and two with peanut yellow lid).

1987 Round Yesteryear Town Scene tins, plain and peanut, 13 and 24 oz. (two with train engine lid and two with horse pulling sleigh lid).

This very early tin held one pound of "Candy Coated Chocolate." Rare.

1988 and 1989 Round Starry Night tins, plain and peanut, 13 and 24 oz. size. The same tins were offered both years and were exactly the same, except for the dates.

1988 Round Vote '88 and Fireworks Tonight! 24 oz. tins, plain and peanut.

1990 Round 24 oz. 50th Birthday tins, plain and peanut.

1991 Prototype round 24 oz. Seasonal Greetings tins with paper labels and scene on white background. This design was replaced by the same scenes with blue highlights and blue lids in the production run and the name was changed from Seasonal Greetings to Season's Greetings on the lids. Rare

1990 Round 24 oz. M&M's On Vacation tins, plain and peanut.

1991 Season's Greetings round 24 oz. tins with Christmas shopping scene in plain and peanut, plus rectangular Holidays Slumber Time tin (plain) and Making A Snowman tin (peanut).

1993 Round light brown (plain) and dark brown (peanut) holiday tins.

1992 Round holiday winter scene (red tin) and holiday snowman scene (green tin).

1994 Round 24 oz. tin with Santa's Workshop (plain), round 24 oz. tin with Santa cooking (peanut), and rectangular green tin with Red baking.

1994 gave us this tin bucket with lid, oval yellow tin, round blue tin, and round brown tin.

1995 blue rectangular Bed & Breakfast tin and beige rectangular Post Office tin.

1996 rectangular Toy Shop tin and rectangular Diner tin.

1997 Octagonal Fire Station tin and octagonal Merry go Round tin.

1998 Hexagonal Sweet Shop tin and rectangular theater scene tin.

Tins & Containers - Miscellaneous

1990s 21 oz. Mini Baking Bits Cookie tins (Double Chocolate Chip Cookies). One is covered in cookies, while the other has the characters appearing on the tin and the lid. $10-20

1990s Special Issue "Save the Children" 32 oz. blue tin with blue plush toy and candy inside. $30-60

990 green cardboard oval (peanut), red cardboard oval (plain - not pictured), 991 green-lidded Elves cardboard oval, and 1989 cardboard oval M&M's esign (plain) containers. Each $20-30

1990s Special Issue "Save the Children" 32 oz. red tin with red plush toy and candy inside. $30-60

989 cardboard oval M&M's design (peanut) container (see above picture for matching (plain) container in the set). $20-30

1991 cardboard oval Easter (peanut) container (matching container for plain candies is unknown). $20-30

1990s M&M's round Santa cookie tin and M&M's square Santa brownie tin. $15-25

1990 round World Cup Soccer cardboard container from Europe. $20-30

1998 7" tall 6-sided NATCO cookie tin. $10-15

Yellow rectangular M&M's Peanut tin. $10-20

1990s William Scotsman Tin. The William Scotsman Company used a modified version of the M&M's logo to advertise their product. Rare

1998 round Big Red cookie tin and Big Red with Yellow cookie tin. $10-15

Toppers

Collectors are frequently confused by the configuration of the base (round or square) and the origin of toppers in their collection.

US toppers have been around since 1988, when the first one appeared for the Christmas season, perched atop a round cardboard tube of plain or peanut M&M's. The first topper came in plain and peanut in orange, red, yellow, and green and had his arms extended out to either side. New, 2 1/2"-3" plastic figures have been a feature of each Christmas, Easter, and Valentine's season ever since in the US. All have round bases and come atop round tubes.

In 1992, topper figures participating in seven different sports were offered. Each sport was represented by a plain and peanut character, but in only one color each (orange weightlifters and skaters, yellow soccer players and skiers, green hurdlers and hockey players, plus a red torch bearer). All of these were issued in the US on square bases on square tubes with the Olympic logo on the tube.

Any topper you find that does not meet the above round/square base criteria is a topper issued outside the US.

Toppers issued in various other countries usually came on square bases, but not always. Some of the non-US toppers are just like their US counterparts (Valentines and Easter toppers are always the same, for example, except that the non-US versions have square bases). Christmas toppers are sometimes slightly different than the US ones. For example, US sledders have their legs apart, non-US sledders have their feet together. The M&M chimney figures have different sacks of toys. Some of the non-US snowball throwers have a pile of snowballs in front of them, some do not. Other square based Christmas toppers are the same as their US counterparts (such as the ones holding candy canes, skaters, skiers, and the ones with their arms extended). A few of the Christmas toppers issued outside the US were never issued in the US at all, such as the ones holding gift boxes or the terrific set of twelve issued in Belgium in 1998.

There were a few Olympic toppers issued outside the US. Eight of them have square bases and are identical to the US ones, except that they were issued in different colors than the US ones. (yellow peanut hockey player, yellow plain skater, red peanut hurdler, red plain hockey player, green peanut weightlifter, green plain torch bearer, brown peanut soccer player, and brown peanut skater without a hat).

Now, here is where it gets interesting. At least five Olympic figures have been found with the same round base as most US toppers have, but these were never available in the US (green peanut weightlifter, yellow peanut torch bearer, red plain torch bearer, brown peanut soccer player, and yellow plain soccer player.) These are extremely rare and hard to find.

Toppers were also issued outside the US for the World Cup Soccer Tournament. All of these have a soccer ball as part of the figure. There are ones with trophies above the figure's head, one with trophies on the base in front of the figure, ones wearing headbands, ones wearing shirts and ones carrying flags. These all had limited distribution and are among the hardest to obtain, especially for US collectors. All of them came in three or four colors and all have square bases, except for the ones wearing shirts.

A sports series of four figures was issued outside the US (tennis player, surfer, snorkler, and motorbiker.) The tennis player has been found in two colors and the others in three colors each. However, it appears that they were released in different colors in different countries at different times. These all have square bases and are also very desirable.

In 1997 and 1998, two different sets of ten toppers were issued in Europe. One is referred to as the occupation series (pilot, clown, chef, etc.) and the other as the sports series (skydiver, in-line skater, skier, etc.) These also have square bases.

There are also some early "double round base" toppers made outside the US. The bases on these are twice as thick as the US toppers. Because they are always found in the very earliest figures, it is assumed that they are the earliest toppers produced outside the US. They appear with the skater, snowball thrower, and arms extended characters. As of this writing, at least eighteen different ones are known to exist with the double round base, including the four soccer players with shirts mentioned above.

To make this area even more interesting, there are some variations—skiers with red pole rings instead of black and Valentine cupid figures with different eyebrows and eyes pointing in different directions, for example.

Following is a listing of all known toppers and variations at this time. The authors hope that you will let us know of any you might have that do not appear in this listing.

ROUND BASE TOPPERS - USA

Christmas Holiday Season

1988

☐	Hands in air with cords - Plain Orange	$3-5
☐	Hands in air with cords - Plain Red	$3-5
☐	Hands in air with cords - Plain Yellow	$3-5
☐	Hands in air with cords - Plain Green	$3-5
☐	Hands in air with cords - Peanut Orange	$3-5
☐	Hands in air with cords - Peanut Red	$3-5
☐	Hands in air with cords - Peanut Yellow	$3-5
☐	Hands in air with cords - Peanut Green	$3-5

1989

☐	On ice skates with cords - Plain Orange	$3-5
☐	On ice skates with cords - Plain Red	$3-5
☐	On ice skates with cords - Plain Yellow	$3-5
☐	On ice skates with cords - Plain Green	$3-5
☐	On ice skates with cords - Peanut Orange	$3-5
☐	On ice skates with cords - Peanut Red	$3-5
☐	On ice skates with cords - Peanut Yellow	$3-5
☐	On ice skates with cords - Peanut Green	$3-5

1990

☐	Snowball with pile in front with cords - Plain Orange	$3-5
☐	Snowball with pile in front with cords - Plain Red	$3-5
☐	Snowball with pile in front with cords - Plain Yellow	$3-5
☐	Snowball with pile in front with cords - Plain Green	$3-5
☐	Snowball with pile in front with cords - Peanut Orange	$3-5
☐	Snowball with pile in front with cords - Peanut Red	$3-5
☐	Snowball with pile in front with cords - Peanut Yellow	$3-5
☐	Snowball with pile in front with cords - Peanut Green	$3-5

1991

(Most of these came with black pole rings at the bottom of the ski pole, but red variations are known to exist for almost all colors, both plain and peanut)

- ❏ Skier with cords - Plain Orange (black pole rings) $3-5
- ❏ Skier with cords - Plain Orange (red pole rings) $15-30
- ❏ Skier with cords - Plain Red (black pole rings) $3-5
- ❏ Skier with cords - Plain Red (red pole rings) $15-30
- ❏ Skier with cords - Plain Yellow (black pole rings) $3-5
- ❏ Skier with cords - Plain Yellow (red pole rings) $15-30
- ❏ Skier with cords - Plain Green (black pole rings) $3-5
- ❏ Skier with cords - Plain Green (red pole rings) $15-30
- ❏ Skier with cords - Plain Brown (black pole rings) $4-6
- ❏ Skier with cords - Plain Brown (red pole rings) $15-30
- ❏ Skier with cords - Peanut Orange (black pole rings) $3-5
- ❏ Skier with cords - Peanut Orange (red pole rings) $15-30
- ❏ Skier with cords - Peanut Red (black pole rings) $3-5
- ❏ Skier with cords - Peanut Red (red pole rings) $15-30
- ❏ Skier with cords - Peanut Yellow (black pole rings) $3-5
- ❏ Skier with cords - Peanut Yellow (red pole rings) $15-30
- ❏ Skier with cords - Peanut Green (black pole rings) $3-5
- ❏ Skier with cords - Peanut Green (red pole rings) $15-30
- ❏ Skier with cords - Peanut Brown (black pole rings) $4-6
- ❏ Skier with cords - Peanut Brown (red pole rings) $15-30

1992

- ❏ Holding candy cane with cords - Plain Orange $4-6
- ❏ Holding candy cane with cords - Plain Red $4-6
- ❏ Holding candy cane with cords - Plain Yellow $4-6
- ❏ Holding candy cane with cords - Plain Green $4-6
- ❏ Holding candy cane with cords - Plain Brown $5-7
- ❏ Holding candy cane with cords - Peanut Orange $4-6
- ❏ Holding candy cane with cords - Peanut Red $4-6
- ❏ Holding candy cane with cords - Peanut Yellow $4-6
- ❏ Holding candy cane with cords - Peanut Green $4-6
- ❏ Holding candy cane with cords - Peanut Brown $5-7

1993, 1994

- ❏ In chimney with toy sack - Plain Red — $3-5
- ❏ In chimney with toy sack - Plain Green — $3-5

- ❏ On sled with legs apart - Peanut Red — $3-5
- ❏ On sled with legs apart - Peanut Green — $3-5

1995

- ❏ Holding songbook - Peanut Green — $3-5
- ❏ Holding songbook - Peanut Red — $3-5

- ❏ Train set around base - Plain Red — $3-5
- ❏ Train set around base - Plain Green — $3-5

1996

- ❏ With snowball and striped hat - Peanut Green — $2-4

- ❏ Riding snowboard - Plain Red — $2-4

1997

- ❏ Elf outfit painting train - Plain Red — $2-4

- ❏ Elf outfit with toy sack - Peanut Green — $2-4

998

❏ Female posing - Plain Green $2-4

❏ Coming out of gift box - Peanut Yellow $2-4

1999

❏ Sitting on gift holding candy cane - Peanut Blue $2-4

❏ Holding on to North Pole - Red Plain $2-4

---

aster Holiday Season

1993

❏ Holding egg with basket at feet - Plain Orange $3-5
❏ Holding egg with basket at feet - Plain Red $3-5
❏ Holding egg with basket at feet - Plain Yellow $3-5
❏ Holding egg with basket at feet - Plain Green $3-5
❏ Holding egg with basket at feet - Plain Brown $4-6
❏ Holding egg with basket at feet - Peanut Orange $3-5
❏ Holding egg with basket at feet - Peanut Red $3-5
❏ Holding egg with basket at feet - Peanut Yellow $3-5
❏ Holding egg with basket at feet - Peanut Green $3-5
❏ Holding egg with basket at feet - Peanut Brown $4-6

1994

- ❏ Sitting in split egg shell - Plain Pastel Green $3-5
- ❏ Sitting in split egg shell - Plain Pastel Pink $3-5
- ❏ Sitting in split egg shell - Plain Pastel Purple $3-5
- ❏ Sitting in split egg shell - Plain Pastel Blue $3-5

1999

- ❏ Wearing purple hat - Plain Green $2-3
- ❏ Lounging in eggshell - Peanut Blue $2-3

Valentine's Day

1995, 1996

- ❏ Holding chick in egg & paintbrush - Peanut Pastel Green $3-5
- ❏ Holding chick in egg & paintbrush - Peanut Pastel Pink $3-5
- ❏ Holding chick in egg & paintbrush - Peanut Pastel Purple $3-5
- ❏ Holding chick in egg & paintbrush - Peanut Pastel Blue $3-5

1997. 1998

- ❏ Holding watering can - Plain Pastel Green $3-5
- ❏ Holding watering can - Plain Pastel Pink $3-5
- ❏ Holding watering can - Plain Pastel Purple $3-5
- ❏ Holding watering can - Plain Pastel Blue $3-5

1992, 1993 (Eyes look to side with straight eyebrows)

- ❏ Cupid with bow & arrow - Plain Orange $3-5
- ❏ Cupid with bow & arrow - Plain Red $3-5
- ❏ Cupid with bow & arrow - Plain Yellow $3-5
- ❏ Cupid with bow & arrow - Plain Green $3-5
- ❏ Cupid with bow & arrow - Plain Brown $4-6
- ❏ Cupid with bow & arrow - Peanut Orange $3-5
- ❏ Cupid with bow & arrow - Peanut Red $3-5
- ❏ Cupid with bow & arrow - Peanut Yellow $3-5
- ❏ Cupid with bow & arrow - Peanut Green $3-5
- ❏ Cupid with bow & arrow - Peanut Brown $4-6

1995, 1996

- ❏ Holding heart and letter - Plain Red — $3-5
- ❏ Holding heart and letter - Plain Pink — $4-6

1994 (Eyes look down with curved eyebrows)

- ❏ Cupid with bow & arrow - Peanut Red — $3-5
- ❏ Cupid with bow & arrow - Peanut Pink — $3-5
- ❏ Cupid with bow & arrow - Peanut Red — $3-5
- ❏ Cupid with bow & arrow - Peanut Pink — $3-5

1997, 1998

- ❏ Holding "I love you" hearts - Plain Red — $3-5

SQUARE BASE TOPPERS - USA

1992 Olympics

- ❏ Torch bearer - Plain Red — $4-6
- ❏ Torch bearer - Peanut Red — $4-6

- ❏ Weightlifter - Plain Orange — $4-6
- ❏ Weightlifter - Peanut Orange — $4-6

- ❏ Skater - Plain Orange — $4-6
- ❏ Skater - Peanut Orange — $4-6

- ❏ Skier - Plain Yellow — $4-6
- ❏ Skier - Peanut Yellow — $4-6

- ❏ Soccer player - Plain Yellow — $4-6
- ❏ Soccer player - Peanut Yellow — $4-6

- ❏ Hockey player - Plain Green — $4-6
- ❏ Hockey player - Peanut Green — $4-6

- ❏ Hurdler - Plain Green — $4-6
- ❏ Hurdler - Peanut Green — $4-6

SQUARE BASE TOPPERS - INTERNATIONAL

Christmas Holiday Season

- ❏ Hands in air with cords - Plain Orange (unverified) — $15-25
- ❏ Hands in air with cords - Plain Red — $6-12
- ❏ Hands in air with cords - Plain Yellow — $6-12
- ❏ Hands in air with cords - Plain Green — $6-12
- ❏ Hands in air with cords - Plain Brown — $6-12
- ❏ Hands in air with cords - Peanut Orange (unverified) — $25-40
- ❏ Hands in air with cords - Peanut Red — $6-12
- ❏ Hands in air with cords - Peanut Yellow — $6-12
- ❏ Hands in air with cords - Peanut Green — $6-12
- ❏ Hands in air with cords - Peanut Brown — $6-12

❏ Snowball - no pile - Plain Red	$10-20
❏ Snowball - no pile - Peanut Yellow	$15-30
❏ Snowball - no pile - Peanut Green	$10-20

❏ On ice skates with cords - Plain Orange (unverified)	$25-40
❏ On ice skates with cords - Plain Red	$6-12
❏ On ice skates with cords - Plain Yellow	$6-12
❏ On ice skates with cords - Plain Green	$6-12
❏ On ice skates with cords - Plain Brown	$6-12
❏ On ice skates with cords - Peanut Orange (unverified)	$15-25
❏ On ice skates with cords - Peanut Red	$6-12
❏ On ice skates with cords - Peanut Yellow	$6-12
❏ On ice skates with cords - Peanut Green	$6-12
❏ On ice skates with cords - Peanut Brown	$6-12

❏ Skier with cords - Plain Yellow	$6-12
❏ Skier with cords - Peanut Green	$6-12

❏ Snowball with pile in front with cords - Plain Red	$6-12
❏ Snowball with pile in front with cords - Plain Yellow	$6-12
❏ Snowball with pile in front with cords - Plain Green	$6-12
❏ Snowball with pile in front with cords - Plain Brown	$6-12
❏ Snowball with pile in front with cords - Peanut Red	$6-12
❏ Snowball with pile in front with cords - Peanut Yellow	$6-12
❏ Snowball with pile in front with cords - Peanut Green	$6-12
❏ Snowball with pile in front with cords - Peanut Brown	$6-12

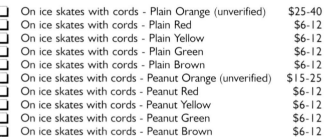

❏ Holding candy cane with cords - Plain Green	$6-12
❏ Holding candy cane with cords - Peanut Yellow	$6-12

❏	In chimney with toy sack - Plain Orange	$6-12
❏	In chimney with toy sack - Plain Red	$6-12
❏	In chimney with toy sack - Plain Yellow	$6-12
❏	In chimney with toy sack - Plain Green	$6-12

❏	On sled with legs together - Peanut Orange (unverified)	$25-40
❏	On sled with legs together - Peanut Red	$6-12
❏	On sled with legs together - Peanut Yellow	$6-12
❏	On sled with legs together - Peanut Green	$6-12

❏	Holding train with another on base - Peanut Red	$10-15
❏	Holding train with another on base - Peanut Yellow	$10-15
❏	Holding train with another on base - Peanut Green	$10-15

❏ Holding train with no train on base - Peanut Red $25-40
❏ Holding train with no train on base - Peanut Green $25-40

❏ Holding gift wrapped box - none on base - Peanut Red $6-12
❏ Holding gift wrapped box - none on base - Plain Orange $6-12

❏ Holding gift wrapped box with another on base - Plain Orange
      $6-12
❏ Holding gift wrapped box with another on base - Plain Red
      $6-12
❏ Holding gift wrapped box with another on base - Plain Yellow
      $6-12
❏ Holding gift wrapped box with another on base - Plain Green
      $6-12

❏ Elf outfit painting train - Plain Red $4-8
❏ Elf outfit with toy sack - Peanut Green $4-8
❏ Female posing - Plain Green $4-8
❏ Coming out of gift box - Peanut Yellow $4-8

❏	In sleigh with reindeer antlers - Plain Red	$10-20
❏	Topping Christmas tree - Plain Red	$10-20
❏	Holding three small gift boxes - Plain Red	$10-20
❏	Holding lamp - Plain Green	$10-20
❏	With three large gift boxes - Plain Green	$10-20
❏	As angel - Plain Green	$10-20
❏	Throwing snowball with tongue out - Peanut Yellow	$10-20
❏	Singing with star and eyes closed - Peanut Yellow	$10-20
❏	With snowman - Peanut Orange	$10-20
❏	Skating and wearing earmuffs - Peanut Orange	$10-20
❏	Carrying sack and holding bell - Peanut Brown	$10-20
❏	On yellow skateboard - Peanut Blue	$10-20

Easter Holiday Season

❑ Holding egg with basket at feet - Plain Red	$6-12
❑ Holding egg with basket at feet - Plain Yellow	$6-12
❑ Holding egg with basket at feet - Plain Green	$6-12
❑ Holding egg with basket at feet - Peanut Red	$6-12
❑ Holding egg with basket at feet - Peanut Yellow	$6-12
❑ Holding egg with basket at feet - Peanut Green	$6-12

❑ Holding rabbit with egg on base - Plain Red	$10-15
❑ Holding rabbit with egg on base - Plain Yellow	$10-15
❑ Holding rabbit with egg on base - Plain Green	$10-15
❑ Holding rabbit with egg on base - Peanut Red	$10-15
❑ Holding rabbit with egg on base - Peanut Yellow	$10-15
❑ Holding rabbit with egg on base - Peanut Green	$10-15

❑ Sitting in split egg shell - Plain Pastel Blue	$15-25
❑ Sitting in split egg shell - Plain Pastel Pink	$15-25

❑ Holding rabbit, no egg - Plain Yellow	$6-12
❑ Holding rabbit, no egg - Peanut Yellow	$6-12

❑ Holding egg - Plain Green	$10-15
❑ Holding egg - Peanut Red	$15-25
❑ Holding egg - Peanut Yellow	$15-25
❑ Holding egg - Peanut Green	$10-15

Valentine's Day

Note eye and eyebrow differences.

(Eyes look to side with straight eyebrows)
❑	Cupid with bow & arrow - Plain Red	$10-15
❑	Cupid with bow & arrow - Plain Yellow	$10-15
❑	Cupid with bow & arrow - Plain Green (not pictured)	$10-15
❑	Cupid with bow & arrow - Peanut Red	$10-15
❑	Cupid with bow & arrow - Peanut Yellow (unverified)	$25-40
❑	Cupid with bow & arrow - Peanut Green (unverified)	$25-40

(Eyes look down with curved eyebrows)
❑	Cupid with bow & arrow - Plain Red	$15-25
❑	Cupid with bow & arrow - Plain Yellow	$15-25
❑	Cupid with bow & arrow - Plain Green	$15-25
❑	Cupid with bow & arrow - Peanut Red	$10-15
❑	Cupid with bow & arrow - Peanut Yellow	$10-15
❑	Cupid with bow & arrow - Peanut Green	$10-15

❑	Holding heart and letter - Plain Red	$60-100
❑	Holding heart and letter - Plain Pink (unverified)	$60-100

1992 Olympics

❏ Ball overhead but no trophy on base - Plain Red	$60-100
❏ Ball overhead but no trophy on base - Plain Green	$60-100

❏ Torch bearer - Plain Green	$8-12
❏ Torch bearer - Peanut Yellow	$8-12
❏ Weightlifter - Peanut Green	$8-12
❏ Skater - Plain Yellow (not pictured)	$10-15
❏ Skater - Peanut Brown (no hat)	$8-12
❏ Soccer player - Peanut Brown	$8-12
❏ Hockey player - Plain Red	$8-12
❏ Hockey player - Peanut Yellow	$8-12
❏ Hurdler - Peanut Red	$8-12

Soccer (World Cup)

❏ Ball in right hand but no trophy on base - Peanut Red	$60-100
❏ Ball in right hand but no trophy on base - Peanut Green	$60-100

❏ Ball overhead and trophy on base - Plain Orange	$15-25
❏ Ball overhead and trophy on base - Plain Red	$15-25
❏ Ball overhead and trophy on base - Plain Yellow	$15-25
❏ Ball overhead and trophy on base - Plain Green	$15-25
❏ Ball in right hand and trophy on base - Peanut Orange (unverified)	
	$40-60
❏ Ball in right hand and trophy on base - Peanut Red	$15-25
❏ Ball in right hand and trophy on base - Peanut Yellow	$15-25
❏ Ball in right hand and trophy on base - Peanut Green	$15-25

❏ Soccer with headband - Peanut Red	$15-25
❏ Soccer with headband - Peanut Yellow	$15-25
❏ Soccer with headband - Peanut Green	$15-25

❏	Wearing shirt - Plain Red	$25-50
❏	Wearing shirt - Plain Yellow	$25-50
❏	Wearing shirt - Plain Green	$25-50
❏	Wearing shirt - Plain Brown	$25-50

❏	Flag - Belgian - Plain Orange	$75-125
❏	Flag - Belgian - Plain Red	$75-125
❏	Flag - Belgian - Plain Yellow	$75-125
❏	Flag - Belgian - Plain Green	$75-125
❏	Flag - Spanish - Plain Orange	$75-125
❏	Flag - Spanish - Plain Red	$75-125
❏	Flag - Spanish - Plain Yellow	$75-125
❏	Flag - Spanish - Plain Green	$75-125
❏	Flag - German - Plain Orange	$75-125
❏	Flag - German - Plain Red	$75-125
❏	Flag - German - Plain Yellow	$75-125
❏	Flag - German - Plain Green	$75-125
❏	Flag - UK - Plain Orange	$75-125
❏	Flag - UK - Plain Red	$75-125
❏	Flag - UK - Plain Yellow	$75-125
❏	Flag - UK - Plain Green	$75-125
❏	Flag - French - Plain Orange	$75-125
❏	Flag - French - Plain Red	$75-125
❏	Flag - French - Plain Yellow	$75-125
❏	Flag - French - Plain Green	$75-125
❏	Flag - US - Plain Orange	$75-125
❏	Flag - US - Plain Red	$75-125
❏	Flag - US - Plain Yellow	$75-125
❏	Flag - US - Plain Green	$75-125
❏	Flag - Italian - Plain Orange	$75-125
❏	Flag - Italian - Plain Red	$75-125
❏	Flag - Italian - Plain Yellow	$75-125
❏	Flag - Italian - Plain Green	$75-125
❏	Flag - Holland - Plain Orange	$75-125
❏	Flag - Holland - Plain Red	$75-125
❏	Flag - Holland - Plain Yellow	$75-125
❏	Flag - Holland - Plain Green	$75-125

Sports

❑ Tennis Player - Peanut Red	$25-50
❑ Tennis Player - Peanut Yellow	$25-50

❑ Biker - Plain Orange	$15-25
❑ Biker - Plain Red	$15-25
❑ Biker - Plain Green	$60-100

❑ Scuba diver - Plain Red	$15-25
❑ Scuba diver - Plain Yellow	$60-100
❑ Scuba diver - Plain Green	$15-25
❑ Surfer - Peanut Orange (not pictured)	$60-100
❑ Surfer - Peanut Red	$15-25
❑ Surfer - Peanut Green	$15-25

Occupations

❑ Fireman - Plain Orange	$6-12	❑ Gardener - Peanut Red	$6-12
❑ Engineer - Plain Red	$6-12	❑ Painter - Peanut Red	$6-12
❑ Sailor - Plain Red	$6-12	❑ Photographer - Peanut Yellow	$6-12
❑ Pilot - Plain Yellow	$6-12	❑ Clown - Peanut Yellow	$6-12
❑ Doctor - Plain Green	$6-12	❑ Chef - Peanut Green	$6-12

❑ Businessman - Plain Red	$10-15
❑ Boat Captain - Plain Red	$10-15
❑ Gondolier - Plain Yellow	$10-15
❑ In-Line Skater - Plain Green	$10-15
❑ Orange raft - Plain Green	$10-15
❑ Photographer - Peanut Orange	$10-15
❑ Skydiver - Peanut Orange	$10-15
❑ Snowshoes - Peanut Yellow	$10-15
❑ Climber - Peanut Blue	$10-15
❑ Astronaut - Peanut Blue	$10-15

ROUND BASE TOPPERS - INTERNATIONAL

Double thickness round bases

❑ On ice skates with cords - Plain Red	$25-50
❑ On ice skates with cords - Plain Yellow	$25-50
❑ On ice skates with cords - Plain Green	$25-50

Double thickness round bases

❑ Hands in air with cords - Plain Red	$25-50
❑ Hands in air with cords - Plain Yellow (not pictured)	$25-50
❑ Hands in air with cords - Plain Green	$25-50
❑ Hands in air with cords - Plain Brown	$25-50
❑ Hands in air with cords - Peanut Red	$25-50
❑ Hands in air with cords - Peanut Yellow	$25-50
❑ Hands in air with cords - Peanut Green	$25-50
❑ Hands in air with cords - Peanut Brown	$25-50

Single thickness round bases; also no ® after the "m"

❑ Snowball with pile in front with cords - Plain Red	$20-40
❑ Snowball with pile in front with cords - Plain Green	$20-40
❑ Snowball with pile in front with cords - Plain Brown	$20-40

1992 Olympics

❏	Torch bearer - Plain red	$40-60
❏	Torch bearer - Peanut Yellow	$40-60
❏	Weightlifter - Peanut Green	$40-60
❏	Soccer player - Yellow Plain	$40-60
❏	Soccer Player - Peanut Brown	$40-60

Toys, Games, Sporting Activities

1998 European Lego Set of 3, "Collect All 3 Sets and Create a Giant Craft!" These three sets came packaged individually in these or the yellow boxes with candy samples. Note: this was more of a Mars promotion than purely an M&M's promotion. Each $35-65

1960s or '70s Candy Gun by Hasbro was also an M&M's candy dispenser! The set came with a "Police" badge, holster, and a bag of M&M's candy to fill the gun. This item appears to be an early issued toy candy gun, as this type of toy would probably not be sold today. $75-150

Pictured are both the yellow and blue versions of the hand held 1998 "Travel Friend" computer games, available in Duty Free Airport Shops. $30-55

This Scandinavian x-ographic space card game consists of a dozen cards that show two different pictures, depending on how you hold them. Set $20-35

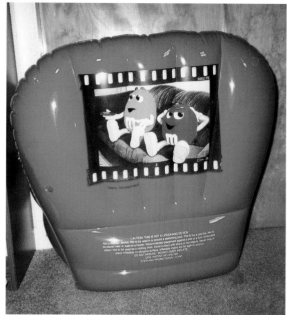

Child and adult sized inflatable vinyl chairs came from the 1998 Employees Catalogue (adult version available later through the Inner Circle™ M&M's Group). Pictured is the child's size chair, front and back.

M&M's Wind Sock.

Kid's cloth pouch in the shape of a pack of M&M's candy holds twenty-six stuffed M&M letters. Rare

1998 Plastic Frisbee, Hackettstown, N.J. (home of M&M's factory) 40th Year Celebration (1958-1998).

1988 Springbok 7"x7", 70 piece jigsaw puzzle. $15-25

European 8" hand puppets came in various colors and were distributed around the world in the early to mid '90s. $20-35

Set of three PENN tennis balls, M&M's figures imprinted on one side. $10-20

1995 Set of six POGS (one duplicate pictured).

Miscellaneous golf accessory items.

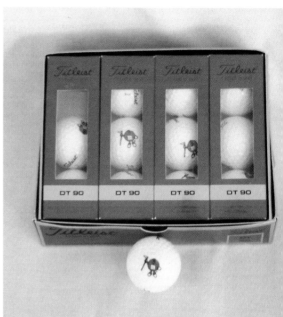

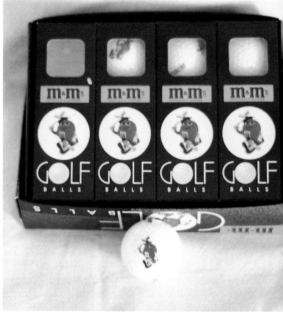

Early M&M/Mars white 10 wheeler truck showing Mars candy products logos, circa 1980s. Truck is larger than today's 1:64 scale. Rare

Matchbox Super Rigs truck, circa 1990s. $10-20

British double-decker bus with Mars advertising.

1996 Matchbox Gold Collection King Size Rigs (1 of 5,000), 1:64 scale. $25-50

Three Yo-Yos from the 1990s!

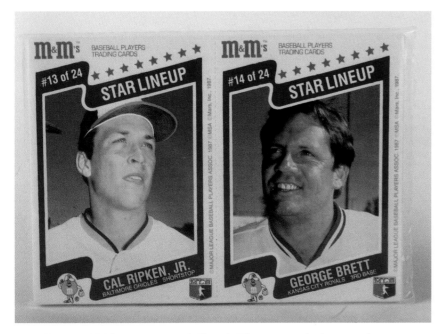

1987 Set of 24 Star Lineup baseball cards. They are attached in pairs, beginning with Cal Ripkin, Jr. and George Brett. $12-25

1984 Set of 44 Olympic Heroes Trading Cards, with box. $20-35

Two Easy-Bake sets showing both packages available, one with M&M's utensils and the other with an M&M canister. $12-25

Older International bead game. Rare

6' Heavy vinyl raft — ideal for family rafting trips! Raft was available only through the Employees catalogue. Rare

Round water game, 4 1/2" diameter. The object is to get free floating M&M's into the stationary M&M's package! Rare

Mars deals a fun hand with these two colorful decks of cards. Other designs have also been offered. $10-15

1999 Rose Art manufactured Domino Cards measures 3 3/4" x 6 1/2" and contain 2 boxes of 28 domino cards. $4-8

Umbrellas

Several different umbrellas have been offered. There is a collapsible black standard-sized umbrella with small characters on the front and back (not pictured), as well as those shown below.

Standard size yellow M&M's umbrella with red, orange, and green M&M's. Rare

Standard size red and white M&M's beach umbrella with blue character - "The BEST chocolate under the SUN!" is marked on the umbrella. Rare

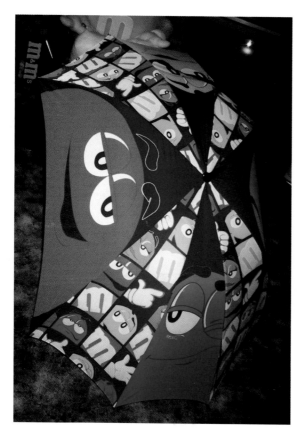

Oblong yellow and brown paneled umbrella, with scene of yellow and red playing golf. $50-75

1999 Oblong umbrella with red, blue, green, and yellow characters divided by design patterned panels. $50-75

Valentines

1980s Package of thirty Valentines! (six each of five designs) Rare

1996 Plastic valentine candy box with Cupid topper figure, plays "Let Me Call You Sweetheart" when you stand Cupid upright. $10-20

Watches

1980s Hamilton M&M's watch, M&M/Mars on the dial.

1996 New Blue Watch. Black plastic case with white background on dial and New BLUE character playing the sax. M&M's are on a revolving dial with a black plastic strap. Watch is dated, 1996.

1987 Watch, yellow plastic case with brown background on dial. "M&M's Chocolate Candies" is printed in white on the dial with four M&M's pictured: yellow, green, red, and orange. Strap is two tone, red upper and green lower (first from left in group picture)

1990 50th Birthday Watch, yellow plastic case with white background on dial and red M&M's in center of dial. Red is wearing a party hat and holding noise maker horn while confetti is flying in the background! Dated 1990 at bottom of dial. Hands are yellow with a red second hand, straps are red upper and green lower (second from left in group picture).

1993 Watch, red plastic case with yellow background on dial. "M&M's Chocolate Candies" is printed in brown on the dial with four M&M's pictured: yellow, green, red, and orange. Strap is yellow upper and green lower. Back is dated, 1993.

1994 Cool Moves Watch! Red plastic case with yellow background on dial. Dial pictures a green skateboard ramp, brown hands, "Red" riding the skateboard. Dated on back, 1994. Dark brown plastic strap.

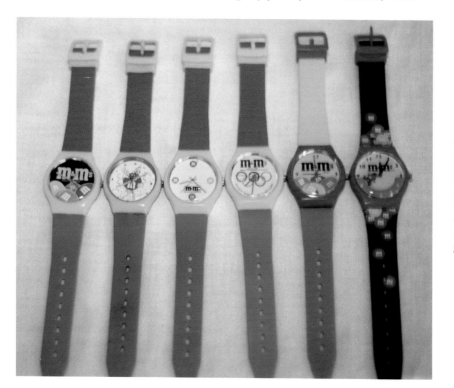

Group photo of watches above plus two additional ones. Third from left: 1992 Watch, yellow plastic case with white background on dial and four M&M's at 3, 6, 9, and 12 o'clock. Hands are yellow with a red second hand, straps are red upper and green lower. Fourth from left: 1992 Olympic Watch, yellow plastic case with white background and Olympic Rings in center. Hands are red, straps are red upper and green lower.

1997 Colorful Character Watch. Black plastic case with white background and "Red M" in center of dial. Watch is dated, 1997. Strap is rubber with characters imprinted on the strap.

1997 M&M's Mini Watch. Case is metal with yellow enamel paint. Hands are red with Minis characters on revolving disk on dial. Strap is purple rubber with Minis characters imprinted.

1998 Millennium Watch, with blue dial and "Red" in the center. Leather strap has "MM means 2000" imprinted on lower strap. "The Official Candy of the Millennium" is imprinted on the inside of the crystal.

1998 European watch in box packaging. Yellow plastic case with white dial and four M&M's at 3, 6, 9, and 12 o'clock. Strap is plastic yellow with minis loose pictured.

Group photo of above watches plus one additional one, shown third from left: Undated, silvertone case with white dial and large "Blue" on center of dial. Strap is black leather. Approximate issue date is late 1990s.

1998 Canadian Minis Watch. Blue and yellow case is metal with white background on dial and "Red" in high top sneakers on the center of the dial. Hands are silvertone and strap is black plastic.

1999 Crispy Edition Watch.

1999 Collectors Edition Racing Team Watch. Silvertone case with black background on dial and "Yellow and Red" characters in the center. "Racing Team" is imprinted on the bottom of the dial. Strap is black leather with "Ernie Irvan #36" imprinted on the upper strap.

Zipper Pulls

1997 "Zipizi" zipper pulls were made out of rubber and stand 1 3/8" tall. $10-15

Index

The Ends